MW00528976

Posthuman Metamorphosis

Posthuman Metamorphosis

Narrative and Systems

Bruce Clarke

FORDHAM UNIVERSITY PRESS
NEW YORK 2008

Copyright © 2008 Fordham University Press

All rights reserved. No part of this publication may be reproduced, stored in a retrieval system, or transmitted in any form or by any means—electronic, mechanical, photocopy, recording, or any other—except for brief quotations in printed reviews, without the prior permission of the publisher.

Library of Congress Cataloging-in-Publication Data

Clarke, Bruce, 1950–
    Posthuman metamorphosis : narrative and systems / Bruce Clarke. —1st ed.
        p.   cm.
    Includes bibliographical references and index.
    ISBN-13: 978-0-8232-2850-8 (cloth : alk. paper)
    ISBN-13: 978-0-8232-2851-5 (pbk. : alk. paper)
    1. Cyborgs in motion pictures.   2. Body, Human, in motion pictures.   3. Cyborgs in literature.   4. Body, Human, in literature.   5. Fantasy literature—20th century—History and criticism.   I. Title.
PN1995.9.C9C53   2008
791.43′656—dc22   2008003066

Printed in the United States of America
10   5 4 3 2
First edition

CONTENTS

ACKNOWLEDGMENTS

At the 1999 Society for Literature and Science conference, Joe Tabbi asked
me to write up Friedrich Kittler's recently translated *Gramophone, Film,
Typewriter* for the *electronic book review*, prompting a rewarding engagement
with media theory. This project began in earnest under the title "Bodies of
Information" with the support of a 2000 Faculty Development Leave from
Texas Tech University. I would also like to thank TTU for its subvention
of my residence in the summer of 2000 at the School of Criticism and Theory at Cornell University. Thanks to David Wellbery and his seminar there,
Observation, Form, Difference: Interdisciplinary Paradigms for Literary
Study, for morphing the project into a prolonged encounter with systems
theory.

I appreciate the opportunities to present parts of the work in progress at
Stanford Humanities Center, Stanford University; the College of Architec-
ture and Planning, Ball State University; Deutsches Haus at New York Uni-
versity; the Departments of English at Texas Tech University, Southern
Methodist University, Wayne State University, the University at Buffalo,
Rice University, and the University of Connecticut at Storrs; the Depart-
ment of Philosophy at Brock University, St. Catherines, Ontario; and the
Department of Languages, Literatures, and Cultures at the University of
South Carolina, Columbia.

Earlier versions of materials incorporated into this study have appeared
as "Mediating *The Fly*: Posthuman Metamorphosis in the 1950s," *Configu-
rations* 10, no. 1 (Winter 2002), published by John Hopkins University
Press; "Paradox and the Form of Metamorphosis: Systems Theory in *A
Midsummer Night's Dream*," *Intertexts* 8, no. 2 (Fall 2004), published by
Texas Tech University Press; "The Metamorphoses of the Quasi-Object:

ix

Narrative, Network, and System in Bruno Latour and *The Island of Dr. Moreau*," in *Revista Canaria de Estudios Ingleses* 50 (April 2005); "Posthuman Metamorphosis: Narrative and Neocybernetics," *Subject Matters* 3, no. 2–4, no. 1 (2007); and "The Self-Referential Scientist: Narrative, Media, and Metamorphosis in Cronenberg's *The Fly*," in *Science Images and Popular Images of the Sciences*, ed. Peter Weingart and Bernd Huppauf (New York: Routledge, 2007).

I am grateful to Len Jenshel (cookjenshel.com) for permission to reproduce the cover graphic, a detail of a photograph taken at Vancouver Aquarium, Canada.

It has been a pleasure to work with my editor, Helen Tartar, and her staff at Fordham University Press. I have also been the beneficiary of manifest forms of aid and assistance from scholarly colleagues. For their generosity and guidance many thanks to Victoria Alexander, Neil Badmington, Laura Beard, Joe Bilello, Linda Brigham, Jim Bono, John Bruni, Tita Chico, Paul Cobley, Hugh Crawford, Katherine Hayles, Mark Hansen, Bernd Huppauf, Minsoo Kang, Edgar Landgraf, Tim Lenoir, Ira Livingston, Lynn Margulis, Robert Markley, Philip Meguire, Colin Milburn, Allen Miller, Hans-Georg Moeller, Tomás Monterrey, Albert Müller, William Paulson, Jim Paxson, Martin Rosenberg, Manuela Rossini, Don Rude, Dorion Sagan, Haun Saussy, Yuan Shu, Joe Tabbi, Hans Turley, Peter Weingart, David Wellbery, Wendy Wheeler, Eric White, Geoff Winthrop-Young, Cary Wolfe, and Karl Zuelke.

And thanks as always to Donna for keeping me afloat.

This work is dedicated to the memory of Octavia E. Butler and Heinz von Foerster.

Posthuman Metamorphosis

Posthuman Metamorphosis

> One might have sought the formation and distribution of the lines, paths, and
> stations, their borders, edges, and forms. But one must write as well of the inter-
> ceptions, of the accidents in the flow along the way between stations—of changes
> and metamorphoses.
>
> —MICHEL SERRES

Metamorphosis

Narratives of bodily metamorphosis depict in various figures the restless
transformations of the human. Over several millennia at least, momentous
corporeal change has been a remarkably stable form of event that connects
the fabulae of mythic and literary narratives. Premodern myth and legend,
folklore and fantasy, set forth the perils of human status by dressing the
sheer contingencies of the natural order in divine or daemonic guises. Scriptural traditions troped bodies and souls into being through spiritual metaphors that attribute human constructions to nonhuman agencies at large in
the extrahuman environment. Against these mythological, theological, and
philosophical backdrops of legendary transformations and eternal essences,
Charles Darwin's *Origin of Species* told an old story in a new way: bodily
metamorphosis is not supernatural but *natural*. Given deep time, biological
evolution is natural metamorphosis. *Origin* narrated species transformations

by human and natural selection, giving scientific legs to a premodern body of fantastic tales that had always implied a greater fluidity of organic embodiment than allowed in Western orthodoxies.

The forms taken by premodern tales of metamorphosis anticipate and overlap modern and contemporary stories of posthuman transformation. But there are some marked differences. Archaic and classical metamorphs—once-human characters mixed up with nonhuman traces—were typically either renaturalized by inscription into the organic order (Daphne into a laurel tree, Narcissus into a flower) or returned to human status after a bestial detour. In contrast, modern depictions of bodily metamorphoses center on reorganizations of biological and/or technological elements without returntickets—uncanny syntheses that retain transgressive stigmata and tend to refuse renaturalization. Such is the fate of the metamorph in Kafka's *Metamorphosis*, in every version of *The Fly*. But, as we will see, this is not always the case with posthuman metamorphs. The metamorphic imaginary since Darwin has a distinctly evolutionary valence: Will this metamorph survive the travail of its transformation, find a way to fit in, to fit circumstances to itself? Can these new formations or mutations achieve viability?

Fictions of bodily transformation are also allegories of new writing technologies; their corporeal variations comprise a cultural history of narrative transformations.[1] The media of social communication are always in material excess of organic contingencies. Metamorphic stories imagine an acceleration of corporeal change in certain living beings, usually attributed to some cultural supplement to the natural order. Amplified by emergent complexities induced by verbal languages and subsequent technologies of communication, social developments outpace biological evolution. Coupled to Darwinian doctrines, modern technological developments have driven the assembly of new stories that rewrite the boundaries between ourselves, animals, and machines. New media generate new metamorphs. Narratives of metamorphosis place the human into improper locations, then try—successfully or not—to renaturalize that impropriety. Whatever the outcome, these stories intimate that the essence of the human is to have no essence.

The Posthuman

In the last two decades the theoretical trope of the posthuman has upped the ante on the notion of the postmodern.[2] The common effect of its several definitions is to relativize the human by coupling it to some other order of being. This study examines a series of narratives of posthuman metamorphosis, culminating in Octavia Butler's Xenogenesis trilogy. From Dr. Moreau's Beast People to David Cronenberg's Brundlefly, Stanislaw Lem's robot constructors in *The Cyberiad* to Butler's human–alien constructs, posthuman metamorphs couple the media systems that enact them to the social systems communicating them to the psychic systems of readers or viewers variously comprehending them. The contemporary discourse of the posthuman signifies a post-Darwinian world, where, as sociologist of technoscience Bruno Latour has remarked, "the human form is as unknown to us as the nonhuman. . . . It is better to speak of *(x)-morphism* instead of becoming indignant when humans are treated as nonhumans or vice versa."[3]

We have noted how current posthuman figures of systemic hybridity have long been anticipated by mythic and literary narratives in which human figures are formally coupled with the nonhuman, transforming them into something beyond the human. In the postmodern moment, the nonhuman other of posthuman couplings is most typically the realm of machines, the "silicon creation" of intelligent circuitry that has arisen alongside the "carbon creation" of evolutionary biology.[4] In *How We Became Posthuman: Virtual Bodies in Cybernetics, Literature, and Informatics*, literary theorist Katherine Hayles posits the philosophical footing of the topic by framing the "human" as the liberal humanist subject constructed by the Enlightenment. Here the subject of the posthuman is precisely the posthuman subject, which, unlike the supposedly universal, natural, and unalloyed Enlightenment subject, "is an amalgam, a collection of heterogeneous components, a material-informatic entity whose boundaries undergo continuous construction and reconstruction."[5] From Hayles's perspective, the virtual bodies imagined or enacted as the material figures of posthuman beings represent the posthuman subjectivities constructed by the coupling of human cognition to digital machinery.

Media theorist Friedrich Kittler rehearses a different prehistory of the posthuman. His discourse in *Gramophone, Film, Typewriter* disperses the humanist subject among media systems.[6] For Kittler, media determine the human situation. The operation of storage and retrieval functions is as important as the contents stored and retrieved. For Kittler, the presumption of spiritual autonomy from such material contexts—the self-determination of the liberal humanist subject—always was a mystification of the relations between persons and communication systems. When writing held a storage monopoly, reading literature—accessing literary narrative in particular as a reservoir of meanings and images—meant learning successfully how to hallucinate the aural and visual implications triggered by bare words on a page. Things have changed since other media began to hallucinate them for us.[7]

These shifts in and around the media environment anticipate others typically identified as cybernetic: the migration of life, sensation, or intelligence from human beings to machines. These cultural events run in parallel with many new chapters in the narrative imagination of bodily metamorphoses. Posthuman metamorphoses indicate how, since the cybernetic convergence of biological and technological notions of evolution, successive models of systems and their environments have played over images of human and non-human bodies.[8] But beyond the vogue for the cyborg and other cybernetic mash-ups of the organic and the mechanical, neocybernetic systems theories illuminate alternative narratives that elicit autopoietic and symbiotic visions of the posthuman. What we stand to gain from these finer discriminations in our contemplation of the posthuman in literature and theory, I will argue, is a more precise appreciation for our evolutionary situation and for the actual complexity of our systemic situatedness.

Systems Theory

The "structural coupling" of different systems and the "interpenetration" of different kinds of systems—these matters lie in the cybernetic domain, and, in this sense, cybernetics is the technoscientific forethought of the contemporary posthuman. Discourses of systems have supervened upon prior idioms of metaphysical souls and genetic identities. The "self-making" formal operations of autonomous living systems—autopoiesis and reproduction—are the literal biological grounds of the cultural figure of bodily metamorphosis. But *metabiotic* systems—psychic, social, and technological systems—are now the cultural contexts through which stories of metamorphosis weave their uncanny interconnections. Since the 1940s cybernetic discourses have attended to the coupling of the human to other orders of things. The idea of the posthuman came of age with the modern coupling of the natural and the technological—matters envisioned at a new level of technical detail in the first cybernetics of Warren McCulloch, Norbert Wiener, John von Neumann, and their colleagues at the Macy conferences on "teleological mechanisms" and "circular causality" in mechanical and social systems.[9]

The discourse of the posthuman has evolved and diverged alongside cybernetics' own mutations. Hayles's posthuman, for instance, is a cautionary trope maintaining an investment in the modern humanist, discursive-dialectical subject. As an ad hoc assemblage of heterogeneous subjectivities aligned to "materialities" of equivocal instantiation, the viability of this posthuman "amalgam" is not clear, but it is probably not intended to be viable. Instead, by means of it, "embodiment" is polemicized and the posthuman rendered as a vitiated informatics having "lost its body." Kittler, in cultural sync with the emergence of cyberpunk culture in the 1980s, seems to enjoy his noir posthuman of reading and writing machines, which render the natural world and all the bodies of its predigital evolution obsolete. In different ways, both Hayles and Kittler frame the posthuman as organic disembodiment. This framing corresponds to first-order cybernetics' insistence on a radical distinction between matter and information, substance and form. In either formulation, we remain in a realm of dialectical antithesis, which observes that the concept of the human has lost its balance and/or its foundations, and that responds either with lament or delight.

True, it is because of the massive accomplishments of the first cybernetics—the technoscientific panoply of communication, control, and computation disciplines—that the coupling of living and mechanical systems saturates the contemporary scene. The emergence of "intelligent" circuitry, micro- and nano-technologies alongside biotechnologies has propelled the mechanomorphosis of the organic body in narrative fiction and in technological fact. Now anthropomorphic figures of fantastic mechanical amalgamation jostle for cultural space. The posthuman remains a heuristic figure for theory construction in the cybernetic mode. But for several decades that mode has diverged into distinct and diverse strands. Pop images of cyborg figures tend to obscure the fact that the posthuman has other significant modalities drawn from these divergent sources. Some lines of development rooted in the cybernetics movement have rebelled against the stereotypes of the prior generation.

Several decades ago Gregory Bateson called for an intellectual transformation with the remark that "the whole of logic would have to be reconstructed for recursiveness."[10] The chief theoreticians and practitioners of this neocybernetic rebellion—Heinz von Foerster, Humberto Maturana, Michel Serres, Francisco Varela, Bruno Latour, and Niklas Luhmann— went to work on the reconstruction project of second-order systems theory. They have consistently stressed that the elemental interdependencies and relational contingencies that produce cognitions of the things we call discursive subjects and scientific objects rely on systems of observation for which no unitary ontological grounding is possible.[11] Von Foerster began putting down the foundations of second-order systems theory in his 1960 paper "On Self-Organizing Systems and Their Environments," which contains seminal discussions of information theory and the system–environment distinction, and anticipates the rise of chaos and complexity sciences with the "order-from-noise" principle.[12] Extending Henri Atlan's bioinformatic concept of "complexity from noise," a direct adaptation of von Foerster, Serres's studies of literature, philosophy, and the history of science show parallel interests in the informatic concept of noise and its role in self-organizing systems.[13] Maturana and Varela make the crucial transition from the self-organization of structures to biological autopoiesis, of which the self-referential operational autonomy of the living cell is the basic example. Directly in line with von Foerster's papers such as "Objects: Tokens for (Eigen-)Behaviors" and "The Cybernetics of Epistemology" and in parallel with their concept of autopoiesis, Maturana and Varela develop a discourse of cognition based on the operational closure of autopoietic observing systems.[14] Von Foerster joins and Luhmann develops their work on the ways that difference and closure among living and nonliving (biotic and metabiotic) systems shape their evolutionary interactions with complex environments.[15]

In second-order systems theory, cognition also undergoes a posthuman metamorphosis. "Subjects" and "objects," observers and the things observed, are cocreating products of self-referential processes. Operational closure is a fact of life, mind, and communication that cannot be wished or scoffed away. The only ground for epistemology is one that is constructed on the fly at the location of, and from moment to moment by, the present operations of observing systems as they observe each other observing. As sociological systems theorist Luhmann has written with regard to the constructivist epistemology of second-order cybernetics: "'constructivism' is a completely new theory of knowledge, a posthumanistic one. . . . The concept 'man' (in the singular!), as a designation of the bearer and guarantor of the unity of knowledge, must be renounced. The reality of cognition is to be found in the current operations of the various autopoietic systems."[16]

Narrative and Neocybernetics

Neocybernetic systems theory resonates with narrative forms, valorizing narrativity as a significant allegory of systemic operations. The temporal operations by which autopoietic observing systems sustain themselves are in no way "trivial"—predictable, reliable, or monotonous. These theoretical tales tell of "nontrivial machines"—metabiotic psychic, technological, and social systems that are just as precarious and unpredictable as those biotic systems on which they rest, the lives of living things.[17] Systems have tales to tell because they have to tell tales—literally, they must sequentially select and connect the elements of a medium in a continuously viable way—to keep going. Narratives of posthuman metamorphosis turn literally and figuratively on matters of autopoietic (dis)continuation.

Chapter 1, "Narrative and Systems," opens with a reflection on narratives of modernity in Jean-François Lyotard, Bruno Latour, and Niklas Luhmann, and ends by surveying a set of modern narratives—*The War of the Worlds*, *Childhood's End*, and *Contact*—that imagine the encounter of modern humanity with alien societies. Between these discussions chapter 1 expounds the logic of systems differentiations in Luhmann's theory construction, to give on that basis a neocybernetic critique and reconstruction of narrative theory. Luhmann's theory of autopoietic systems distinguishes their kinds into living, psychic, and social varieties, stressing their operational autonomy but also the continuity of recursive self-reference between biotic (living) and metabiotic (psychic and social) systems. Systems of consciousness and communication produce their differential forms in the shared medium of *meaning*. The operational autonomy and coupled interpenetration of social and psychic systems present a rigorous view of matters often glimpsed but not firmly grasped in previous narratological discussion, in particular, the conceptual autonomy yet discursive merger of narration and focalization. Redescription of mainstream narrative theory and science fictions through second-order systems concepts recuperates their residual humanisms for a posthumanist narratology that factors narrative communications into the wider complex of social and psychic functions.

Chapter 2, "Nonmodern Metamorphosis," discusses transformational dynamics in the work of Bruno Latour. Pursuing a pragmatic rather than systematic form, that work rises to full-blown "scientifictional" narrative in the text of *Aramis, or the Love of Technology*. In that text and in *We Have Never Been Modern*, Latour elaborates an exposition (derived from Michel Serres) of the *quasi-object* as a redescription of the ways that the social networks of technoscience construct their elements. Serres's quasi-object is a thoroughly neocybernetic concept interrogating the ontological foundations of the subject–object dichotomy as an oversimplification of complex systemic couplings—of the "hybrids" constructed by sociotechnical networks. Just as informatic feedback complicates the input–output and signal–noise distinctions, the hybrid couplings of technoscience in action complicate distinctions between cultural subjects and natural objects. These complications are vividly present in a seminal narrative of technoscientific modernity and its discontents, Wells's *The Island of Dr. Moreau*. The misguided Dr. Moreau creates in the Beast People quasi-human quasi-objects that he then discards but fails to repress. Their "post-animality" joins the posthuman aliens examined in chapter 1 in undermining anthropocentric defenses against the nonessentiality and nonmodernity of modern humanity.

The logical, discursive, and epistemological problems of paradox are prime interests of neocybernetic systems theory.[18] Chapter 3, "System and Form," looks at the ways narrative theory dovetails with neocybernetics in the formalisms of paradox. Systems theory factors into narrative at the level of systemic operations and the level of formal structures. To set up our explorations of paradox in rhetorical frames and narrative plots, we will make an introductory pass through George Spencer-Brown's *Laws of Form*, a "calculus of distinctions" to which Luhmann often turns to explain how paradox becomes operational in self-referential systems. Narrative forms such as the plotting of episodes, diegetic levels, and character-bound, authorial, and figural narrative situations are also illuminated at the nexus of neocybernetics and *Laws of Form*. An important distinction in this milieu is that between "first-order" and "second-order" observation, the latter arising recursively from the "observation of observation." As narratives may be construed not just in their narrating operations but also as observing systems, critical observation of narrative observation can utilize second-order formations.

A related paradoxical recursion investigated in chapter 3 is the form of *reentry*, through which an observing system reenters the distinction between itself and its environment into its own operations. Self-referential paradox is also a virtual component of the narrative metamorph, an improper protagonist built on the predication of contradictory, logically incompatible minds and bodies. Playing with fairy lore on the cusp of modernity, Shakespeare's metamorphic comedy *A Midsummer Night's Dream* unfolds and refolds the paradoxes of form and distinction by staging the observation of observation, and in particular the feedback between the observation of transformation and the transformation of observation. On the cusp of postmodernity, in Damon Knight's 1964 science fiction *Beyond the Barrier*, contradictory two-sided forms push along a fable that climaxes in a moment of posthuman metamorphosis.

Chapter 4, "Metamorphosis and Embedding," reads the narrative discursive play of embedding and embedded frames—stories within stories and plays within plays—as a primary literary analog of systemic recursion and self-reference. Stories of metamorphosis typically operate on multiple and shifting narrative levels, and these recursions of formal structure reverberate in the reframed bodies of narrative actors. A number of narrative theorists link the topic of embedded narrative to issues of life and death, a biotic reference that we will redescribe neocybernetically as an allegory of metabiotic autopoiesis. The play at the boundary between different narrations also provides a point of connection between Gregory Bateson's pioneering cybernetic discussion of levels of communication in the "play frame" and Jacques Derrida's discourse of the *parergon*—that which is "beside the work" from pedestals to paratexts. Both discourses instance the paradoxical boundaries systems construct among shifting levels of observation. Crafting its fictive scenarios from the feedback of cybernetic content and cybernetic form, Stanislaw Lem's *The Cyberiad: Fables for the Cybernetic Age* (1967; Eng. trans. 1974) weaves its narrative forms out of embedded levels and paradoxical constructions. Delving deeply into the epistemological and ontological repercussions of cybernetic ideas, *The Cyberiad* anticipates neocybernetic themes by exploring the formal resonances between cybernetic and narrative recursion.

Systems theory emphasizes that system maintenance is system evolution. Metamorphoses from informatic mutations and systemic mergers are inevitable. The traditions remembered in the classical metamorphic stories of

Ovid and Apuleius anticipate this neocybernetic understanding: the reproduction and transmission of any system opens it to metamorphosis. Latour
says it like this: *"no transportation without transformation."*[19] If the first cybernetics focused on the flows of information within, among, and across natural
and mechanical systems, media theory since Marshall McLuhan has been
directed to the technological forms and media structures that carry those
signals from system to system. In second-order cybernetics, social systems
arise through a nested interaction of operationally closed psychic systems:
if structural coupling over spatial distances and temporal gaps is to occur,
media of some sort must intervene.

Chapter 5, "Communicating *The Fly*," shifts the discussion of narrative
embedding toward regimes of *remediation*—the embedding of one media
technology inside another—and picks up the discussion of metamorphosis
in its cybernetic transformation through media and computing devices.
Under transmission and receptive reconstruction, original messages do not
remain the same, but are deformed by noises traversing the media through
which they become materialized and reformed by whatever cognitive apparatus is at hand to (mis)understand the communication. The fantasy of in-
terrupted teleportation that drives each *Fly* is inscribed in larger reflections
on the psychic and social uncertainties of communications media. In each
version of *The Fly*, the fabula turns on a noisy transmission from an oversight by the operator of a teleporter, a machinic hybrid of material transportation and informatic transmission that becomes an allegorical operator in
a self-referential narrative of metamorphic catastrophe. By the time David
Cronenberg directs his 1986 remake of the 1950s versions, the personal-
computerization of the teleporter adds to the text a posthuman narrator that
drives the updated metamorph—the Brundlefly—to the edge of posthuman
viability. But here too a narrative of posthuman metamorphosis falls captive
to humanist revisions that demonize rather than operationalize its self-referential thematics.

In many modern narratives of metamorphosis, then, humanist assumptions stand in the way of posthuman possibilities. It is fairly typical for first-
order cybernetic ideas to be linked with pseudo-evolutionary notions pitting
the "nature" of the human against posthuman monstrosity, and for the destruction of the posthuman metamorph to be read as an affirmation of the
human status quo. But Donna Haraway's classic "Cyborg Manifesto" concludes with an evocation of science fiction, feminist science fiction in particular, as a realm of potentially liberatory reimagination of the human

through the trope of the cyborg.[20] Haraway later extends this point to the narrative component in technoscience generally and in the texts of its critical study: "Fact and fiction, rhetoric and technology, and analysis and storytelling are all held together by a stronger weld than those who eschew taking narrative practice seriously in science—and in all other sorts of 'hard' explanation—will allow."[21] The implication is that affirmative (neo)cybernetic possibilities for the (post)human can be determined, effectively informed by such narrations.

Chapter 6, "Posthuman Viability," concentrates on a major example of affirmative neocybernetic posthumanism: Octavia Butler's Xenogenesis trilogy (1987–89). And written all over that trilogy is the neocybernetic biological discourse of Lynn Margulis and Dorion Sagan.[22] As Margulis and Sagan tell it, the basal form of every living system is the autopoietic, self-referential cellular system. In deep evolutionary terms, life's corporeal metamorphoses begin with the genetic exchanges of the bacteria, and after two billion years of bacterial experimentation and divergence, organic transformations really start to roll with the assemblage of prokaryotic bacterial cells into the eukaryotic or nucleated cells out of which all other, later-emerging forms of life are built. Butler's incorporation into the Xenogenesis trilogy of a separate but symbiotic genetic packet (the "Oankali organelle") that can be embedded into and so transform other cells is modeled on Margulis's discussion of the mitochondrion within her larger theory of symbiogenesis.

The title of the Xenogenesis trilogy's reissue in an omnibus edition, *Lilith's Brood* (2000), underscores the procreative scenario providing this "brood"—the genetic nucleus of a posthuman metamorphic populace—with a chance for reproductive, psychic, and social viability. Underwritten by neocybernetic biological theory, this narrative painstakingly depicts the emergence of a "construct" species in which human biology and society are both included and transcended. In the Xenogenesis trilogy, a plausible form of posthuman viability entails metamorphosis not just instrumentally for the passage into the posthuman, but ultimately as a physiological faculty attainable by the posthuman body itself. The Xenogenesis trilogy raises posthuman metamorphosis to the second and third powers.

Posthuman Metamorphosis: Narrative and Systems reads narratives of bodily metamorphosis as allegories of the contingencies of systems. Its points of literary reference are the posthuman intuitions of both pre- and postcybernetic metamorphs. Its discursive concerns are to demonstrate the viability of second-order systems theories for literary-critical, narratological, and cultural work. The questions it addresses include: How do psychic and social systems use images and narratives of the metamorphic body to conceptualize and communicate? What do past and present bodily forms now mean as observed in an era of cybernetic multiplicity? These are not perhaps the typical questions of literary enquiry, but I hope my reader will find that they are important and appropriate ones to pose in the current cultural situations of technological hybridity, systemic complexity, and social volatility.

ONE

Narrative and Systems

Narratives of Modernity

Narrative is a primary formal and thematic program running on the complex infrastructures of social and psychic systems. The medium of narrative in society is the network of metabiotic meaning systems and their media environments. The maintenance-in-being of narratives in any textual medium has to be continuously reconstructed within social systems that can use them as elements of communicative exchange. Over time these contingencies ensure the continuous transformation of narratives and, from fictions of metamorphosis to histories of social evolution, the continuous recreation of narratives of transformation.

For instance, the emergence of cybernetics has been changing the way we narrate modernity. We have only recently noticed that systems are inexorably coupled to the environments they distinguish themselves from to arise as systems. It is as if the environments of systems had long occupied cognitive blind spots from which they have now been shifted into view. Yet the belated emergence of our environmental awareness seems to have accompanied or occasioned a related breakdown in modern social certainty, particularly Western certainties about the "universality" of its definition of the human and the global centrality or "firstness" of its "world." Jean-François Lyotard famously defined postmodernity as the historical period that witnesses that lapse: "The grand narrative has lost its credibility, regardless of what mode of unification it uses, regardless of whether it is a speculative narrative or a narrative of emancipation."[1] This withering of the grand narratives of modernity has been variously felt to be a loss or a gain, but still, a profound event, worthy of its own historical distinction.

Niklas Luhmann's discourse on modernity acknowledges this description but deflates its profundity. Luhmann withdraws the singularity of this semantics and reinscribes this description as an unavoidable effect of modern social complexity. "Postmodernity" is one of many ways to observe communicative formations in the operation of modern social systems and their dissolution over time. "The proclamation of the 'postmodern' has at least one virtue," he writes in the preface to his *Observations on Modernity*: "It has clarified that contemporary society has lost faith in the correctness of its self-description."[2] But this in itself fails to distinguish the postmodern from the modern per se. Rather, in Luhmann's account, the phenomenon Lyotard has described is a belated acknowledgement of the fictionality of grand narratives all along: "There is no *métarécit* because there are no external observers. Whenever we use communication—and how could it be otherwise—we are already operating within society."[3] That is, even if we could arrive at a metaperspective on the social systems in which we are already interpellated, we could not communicate that perspective to those systems. Just as the mass media co-opts any attempt to use them to indicate a state of affairs transcending the society at hand—"transgression and subversion never get 'on the air' without being subtly negated as they are; transformed into models, neutralized into signs, they are eviscerated of their meaning"[4]—the reception of communication within given systems deletes the metasystemic character of any message.[5]

Bruno Latour has suggested another provocative version of the breakdown of modern metanarratives. In *We Have Never Been Modern*, Latour purports to deconstruct the metanarrative of modernity itself.[6] A summary of Latour's historical thesis might go like this: modernity arose from the suppression of the supposed incoherence of premodern ideas—their mixing up of natural and cultural causalities—but this sorting operation occurred at the price of enforced separations between natural objects and social subjects that have now in turn lost their coherence as well. According to Latour, by pursuing an anthropology of technoscientific practices, critical science studies has now unraveled the modern unraveling of premodernity. Latour would reform academic protocols and resolve the distinctly modern conflict between industrial societies and the ecological constraints of the natural world. He concludes *We Have Never Been Modern* with a prescription for a politics of nonmodern collectives, presenting the state of "nonmodernity," complete with a "nonmodern Constitution" for a "parliament of things," as a way beyond the impasses of modern ontological separatisms.

Luhmann's theory of modern social systems, in contrast, is uninflected by Latour's brand of dialectical drama.[7] What Latour perceives as modernity—an enforced purification covering over an orgy of hybrid copulations—Luhmann views as a change in habits of differentiation and a move from a less to a more complex form of differentiation. We *are* modern, Luhmann would insist, because of an irreversible social evolution he describes in the following manner: social differences in premodern societies were stratified, predetermined by a hierarchical system of proper social positions; modernity arose over time as stratified social differentiation gave way to the differentiation of social functions—for one instance, the separation of church and state. Differentiations of function (rather than of strata) now fissure society as a hierarchical totality into autonomous but interconnected subsystems of government, law, science, religion, education, industry, media, and so forth. Society is now a hypercomplex system of systems within systems, occupying a built environment that is itself a network of material–mechanical systems embedded in natural environments that are often unperceived until, on occasion, they twitch catastrophically, or, now that global warming has crossed an observational threshold, undergo a potentially millennial transformation.

Luhmann would neither predict where society is headed nor prescribe a direction toward which to steer it, since it is in principle unsteerable. But he did specify some constraints on the operations of autopoietic systems. They must constantly submit their elements to dissolution to continue reproducing themselves. The constant reshuffling necessary to systemic self-reproduction ensures the emergence of difference over time: system evolution. This description goes beyond, but includes and valorizes, social democracy in its liberal and libertarian senses as a dynamic equilibrium of social forces and a monitored interrelation of quasi-emancipated institutions. Based on such formal considerations of the abstract conditions of systemic possibility, Luhmann emphasized the paradoxicality and unpredictability that inflect social systems' temporal evolutions.

Metabiotic Systems

In his "Excursus on Luhmann's Appropriation of the Philosophy of the Subject through Systems Theory," Jürgen Habermas notes Luhmann's appropriation of Maturana and Varela's concept of biological autopoiesis for "metabiological" extension to psychic and social systems.[8] He criticizes the propriety of that appropriation not from the side of autopoiesis as a biological theory, but rather from the side of "communicative reason" as a social philosophy. His own theory of communication provides a "concept of communicative reason developed in terms of linguistic functions and validity claims"; with Luhmann, on the contrary, "the metabiological frame of reference does not go beyond the logocentric limitations of metaphysics, transcendental philosophy, and semantics," as he considers his own philosophy to do, "but undermines it. Reason once again becomes a superstructure of life. In this respect, nothing is changed by promoting 'life' to the organizational level of 'meaning.'"[9]

Habermas's *metabiology* would be akin to a garden-variety sociobiology with a hard cybernetic twist, in which genetic or organic or otherwise "systemic" (or inhumanly "machinic") determinisms subsume the functions of rational sense and consensus within the metaphysical "lifeworld." In this paradigmatic misreading, Habermas refuses the posthumanist turn that Luhmann has given to biological autopoiesis. For Luhmann, it is not the case that life has been "promoted . . . to the organizational level of 'meaning,'" in which case such a metabiology would still remain categorically bio-logical. It is rather that Luhmann has abstracted from Maturana and Varela's recursive formalism of biological systems a paradoxical paradigm of operational closure within environmentally open, self-referential systems that he then applies, by reformulating their elements as meaning-events, to the systematicity of psyches and societies. In other words, it would be more accurate, if less elegant, to say that Luhmann's social systems theory is "meta-metabiological." The term we prefer is *neocybernetic*.

> Living systems are a special type of systems. However, if we abstract from life and define autopoiesis as a general form of system-building using self-referential closure, we have to admit that there are non-living autopoietic systems, different modes of autopoietic reproduction, and general principles of autopoietic organization which materialize as life, but also in other modes of circularity and self-reproduction. In other words, if we find non-living autopoietic systems in our world, then and only then will we need a truly general theory of autopoiesis which carefully avoids references which hold true only for living systems.[10]

The posthumanist turn in Luhmann is encapsulated in the fundamental theoretical move of positing *nonliving autopoietic systems*. Folded up in Luhmann's rather flat locution is an important difference in the ways a system can be nonliving—nonliving non-autopoietic systems and nonliving autopoietic systems. The terminology I will use to make the implications of this distinction explicit is *abiotic* for the former and *metabiotic* for the latter. Given this distinction, we can observe how the concept of "the human" is incoherent, in that "the human" is at once both living and nonliving. More precisely, and less paradoxically, the human is at once both biotic and metabiotic.

I will further unfold some of the implications of this set of distinctions through an elaboration of Luhmann's own diagrams. At the beginning of *Social Systems*, Luhmann offers this sketch:

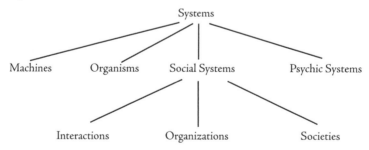

Figure 1.1. Systems Distinctions, from *Social Systems*

The first level of systems distinctions here remains equivocal because organisms, social systems, and psychic systems are variously autopoietic, whereas machines are nonautopoietic. In a related diagram from "The Autopoiesis of Social Systems," that situation is remedied by leaving mechanical systems aside:

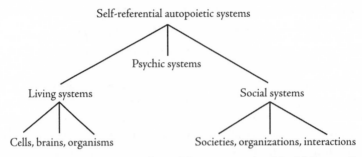

Figure 1.2. Systems Distinctions, from "The Autopoiesis of Social Systems"

But this triadic scheme still leaves unmarked the distinction between living and nonliving autopoietic systems expounded in the accompanying text. Let us rearticulate Luhmann's distinction between living and nonliving autopoietic systems through the distinction of biotic from metabiotic systems:

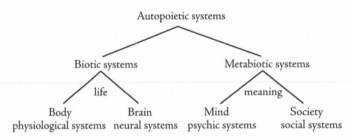

Figure 1.3. Autopoiesis of Biotic and Metabotic Systems

Now let us add in the other half of the distinction we have made in the category of the nonliving—the abiotic versus the metabiotic—along with the other half of the autopoietic–nonautopoietic distinction. This will reintegrate the category of machine systems in a distinct and perhaps surprising way. We see that machines are constitutionally paradoxical—they are nonautopoietic subsystems that emerge out of environments compounded with metabiotic autopoietic systems:

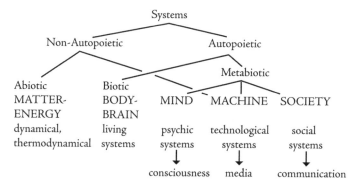

Figure 1.4. Autopietic and Non-Autopoietic Systems

Unfolding Luhmann's neocybernetic interventions, we have constructed another, complementary, and more familiar angle on the concept of the posthuman: technological systems, while nonautopoietic, are nonetheless as emergently metabiotic as psyches and societies, and therefore are a species of meaning systems in their own right.

Figure 1.4 would perhaps be more rigorous if the Machine column were placed to the right of the Society column, out from under the category of metabiotic autopoietic systems. But I have drawn the figure "paradoxically" to indicate the coemergence of technological systems and the roles they have always played as mediating structures connecting psychic and social systems. Writing and narrative belong here, along with cell phones and airplanes. As Latour remarks, "Every machine is scarified, as it were, by a library of traces and schemas"[11]—marks impressed upon them by the psychic and social systems that form the environments out of which technological systems arise, and for which they provide their mediations. Here, once more, the posthuman is something more precise than a mash-up of the human and the mechanical. It is bound up with the further unfolding of a metabiotic emergence of nonautopoietic systems and structures, distinct from, yet coeval with, the coevolution of autopoietic meaning systems.

Posthumanist Narratology

I. DISCOURSE AND STORY

Contemporary narratology dates from the 1960s, when its structuralist progenitors adopted a model of narrative developed in the 1920s by the Russian
Formalists—the *sujet–fabula* distinction—and retooled it into the distinction
between *discourse* and *story*.[12] This two-component structure is now the
foundational model for mainstream narrative theory. *The Cambridge Introduction to Narrative* gives a representative example of how narrative is defined by analysis into these two terms: "Narrative is *the representation of
an event or a series of events*. . . . The difference between events and their
representation is the difference between *story* (the event or sequence of
events) and *narrative discourse* (how the story is conveyed)."[13] But while this
codification of narrative analytical technique marks a decisive improvement
over theoretically naive approaches, narratology continues to struggle with
problems generated when this two-component model oversimplifies the full
narrative phenomenon it describes. A superior model has been developed
and is in circulation, but it still appears to be the exception in the literature
of narrative theory. Derived within narratology proper, this preferable
three-component model anticipates or parallels key systems-theoretical distinctions and can be redescribed in terms of its systems-theoretical
implications.

Gérard Genette's *Narrative Discourse: An Essay on Method* (French original published in 1972) is both hero and villain in this theoretical saga. The
introduction to this brilliant but convoluted study provides a preliminary
template for the three-component model mentioned above, but Genette
himself then goes on to deploy and reinforce the dyadic discourse–story
(sometimes discourse–diegesis) model almost exclusively. But note his
opening remarks. When it comes to "narrative," "We must plainly distinguish under this term *three* distinct notions" (my italics):

> A first meaning . . . has *narrative* refer to the narrative statement, the oral or
> written discourse that undertakes to tell of an event or a series of events.[14]

This is the *discourse* proper, the primary topic of *Narrative Discourse*.

> A second meaning . . . has *narrative* refer to the succession of events, real or
> fictitious, that are the subject of this discourse.[15]

This is the *story*, the presumably causal and chronological *histoire* made available by the discourse. Here already is the full template for the simple discourse–story distinction. But Genette continues:

> A third meaning, apparently the oldest, has *narrative* refer to . . . the event that consists of someone recounting something: the act of narrating taken in itself.[16]

But Genette downplays this third sense by identifying it not with the *narrator*'s but the *author*'s production, which would place it outside the realm of narrative discourse per se. Rather, it is (in the *first* meaning above) "narrative, and that alone, that informs us here both of the events that it recounts and of the activity that supposedly gave birth to it. . . . Story and narrating thus exist for me only by means of the intermediary of the narrative."[17] Only in the final chapter will Genette have recourse to the phrase "narrating instance," to indicate the event of the (literary) narrative text's virtual production by a (real or fictive) voicing—*narrating*—of its signifiers.

Unlike in his discussions of *order, duration, frequency,* or *mood,* to give a comprehensive treatment of *voice—who speaks?*—Genette had to use a combination of conceptual categories—narrative discourse and narrating instance. This circumstance indicated from the outset that his concept of narrative discourse was overdetermined and, to be theoretically functional in a less problematic way, in need of some dedicated unpacking. To my knowledge, the narratologist who has most satisfactorily resolved this issue is Mieke Bal. Clearly Bal cribbed large parts of her system from Genette, but she earned that right with some decisive and positive alterations. Why her three-layer model has not been more widely adopted I cannot say, unless it is the fear of ever-proliferating equivocations in technical jargon struck by "the large Terminological Beastie looming over the field."[18] The only drawback I can locate in Bal's system, a relatively minor one, is not conceptual but terminological: what mainstream narratologists from Genette on call *discourse,* she divides up as *text* and *story*; what they call *story,* she calls (returning to Formalist roots) *fabula.*

My references to Bal's system will be taken from the second, explicitly *post*structuralist edition of her *Narratology.* The introduction to this volume states: "a three-layer distinction—text, story, fabula . . . is the point of departure for the theory of narrative texts that is presented here."[19] In this rearticulation of analytical categories, the term *discourse* is semi-retired and the overdetermined concept that it indicated is separated into *text* and *story.*

Their interrelation is not representational but semiological: viewed from this angle, if discourse altogether denotes the narrative sign, text is the narrative signifier—"a finite, structured whole composed of language signs. . . . Only the text layer . . . is directly accessible"[20]—and story is the signified content of those signifiers. The semiotic rather than literary orientation already allows for what Marie-Laure Ryan will call a *transmedial* narratology.[21] Even though Bal's interests are predominantly literary, her stipulation of the text layer allows more direct attention to differential matters of narrative media: "'text' refers to narratives in any medium. . . . A narrative text is a story that is 'told' in a medium. . . . If one regards the text primarily as the product of the use of a medium, and the fabula primarily as the product of imagination, the story could be regarded as the result of an ordering."[22]

Bal's story layer is a more finely tuned and delimited version of Genette's discourse, in that its "intermediary" status is explicitly specified as an *ordering* that is complex (or two-sided: see chapter 3). Turned toward the text, it reflects how narrative discourse takes perceptible shape in the first place by impressing a specific sequence of semiotic forms into one or more textual media. At the same time, turned toward the fabula, the story layer is an ordering of the event or events to be told. "A *fabula* is a series of logically and chronologically related events," while "a *story* is a fabula that is presented in a certain manner"; or, the story is the manipulation of the fabula; "the fabula is 'treated'" by the story.[23] The key to working Bal's complex analytical system is to keep in mind that all layers are in virtual operation at once and from both directions.[24] On one hand, in a narrative's emergence into and dwelling within discourse as a sharable, duplicable communicative artifact, the (imaginative or other) production of a fabula may induce its treatment in a story that induces its inscription in a text. On the other hand, in its reception by a user, the inscriptions presented by a text produce the cognition of a story that produces, to whatever extent desired or possible, the (re)construction of a fabula.

For the purposes of a posthumanist narratology, one immediate benefit of the tripartite division of Bal's scheme is its effective deconstruction of narrative personhood into an "anthropomorphic figure, called 'actor' in the study of the fabula, 'character' in the study of the story, and 'speaker' in the study of the text."[25] While not equivalent to the disarticulation of systems references typically bundled into the factitious unity of the "human sub-ject," it is at least consonant with it. Let us begin to redescribe these narra-tological matters in systems-theoretical terms. After all, narratives go nowhere and do nothing until they enter and circulate as communicative proposals within social systems. Because social systems carry out their own autopoiesis by producing and reproducing communications, one will need a "concept of communication . . . that strictly avoids any reference to con-sciousness or to life, that is, to other levels of the realization of autopoietic systems."

> Like life and consciousness, communication is also an emergent reality, a self-generated state of affairs. It comes about through a synthesis of three different selections, namely the selection of *information*, the selection of the *utterance [Mit-teilung]* of this information, and selective *understanding or misunderstanding* of this utterance and its information.[26]

Communication:	the selection of:	the selection of:	the selection of:
in Luhmann's social systems-theoretical analysis	information	utterance	(mis)understanding
in mainstream communication theory	information source, sender	message and channel	destination, receiver
as adapted to narrative theory:	events	the representation of the events	the (re)construction and/or interpretation of the events
in Genette et al	(narrating instance)	discourse	story
in Bal	text	story	fabula

Table 1.1. Social and Narrative Communication

For communication to occur, it is not enough just to send or receive a message. Understanding must close the loop by processing the utterance as a meaningful selection of information. If that happens, communication oc-curs and is in a position to continue. In the narratological instance, the operational circularity or recursive functionality key to Luhmann's model of autopoietic systems is already implicit in my statement above that all layers operate at once and from both directions. The construction of a meaningful fabula (understanding) feeds back into the cognition of the story (utterance) presented by the textual signifiers (information). Over the course or temporal duration of a narrative communication, each phase and component of the process has the potential to enhance and reinforce—or, alternatively, to disrupt and break down—the entire complex performance. "The system pulsates, so to speak, with the constant generation of excess and selection."[27]

2. OBSERVING THE MEDIUM

Narrative, media, and systems theories, in their various investigations of the ways media move in and out of blind spots, can compensate for each others' partialities. We are working on a neocybernetic synthesis that refers these cognitive matters to the coupled and contingent autonomies of perception and communication: psychic and social systems. Luhmann brings the concept of *medium* forward at different levels of application. At its most abstract, medium is inscribed in the theory of form, not as its opposite, but as its outside. Medium is the unmarked space, the unobserved plenum of unfixed possibilities upon which the indications of form put their marks. An undifferentiated medium is the condition of the possibility of any form of difference. Form arises as a difference from the indifference of its medium. The emergence of distinctions and forms presupposes the availability of media in the environment of those processes. "I shall characterize media by their higher degree of dissolubility together with the receptive capacity for fixations of shape (*Gestalt*). . . . Air is only a medium because it is loosely connected in this way. It can transmit noises because it does not itself condense to noises. We only hear the clock ticking because the air does not tick."[28]

Air is an apt example of a generalized medium because, when it is at rest, to normal observation it is literally imperceptible—invisible, inaudible, the unmarked and unremarked space of every aural form it carries to an observer. It functions as a medium because it is in fact an elastic colloid of unattached molecules. "Forms by contrast arise through the concentration of relations of dependence between elements, i.e., through selection from the possibilities offered by the medium. The loose connection and easy separation of the elements of the medium explains why the medium is not perceived but only the form that coordinates the elements of the medium."[29] But when the wind howls, the materiality of this medium does indeed "condense to noises," unintended forms that may overwhelm all others in the air—say, cries for help. As we will discuss further in chapter 5, noise is the self-reference of the medium. But most of the time, the near imperceptibility of air as a medium renders it emblematic of the tendency of all media to disappear behind the forms they make possible.

For the processing of forms within psychic and social systems, however, the medium in question is not physically substantial, such as air or water, but entirely virtual—it is *meaning*. Meaning is the self-generated medium of the self-referential self-maintenance (autopoiesis) of minds and societies out of the systemic recursion of their own forms. As the medium of the interpenetration of psychic and social autopoieses, meaning is the air we breathe, out of which this or that meaning, momentarily here and now, comes and goes. "A system that is bound to use meaning as a medium constitutes an endless but complete world in which everything has meaning, in which everything gives many clues for subsequent operations and thereby sustains autopoiesis, the self-reproduction of the system out of its own products."[30] For psychic and social systems, the desire for meaning is constantly regenerated by a world whose meaningfulness is always already constructed by the forms of consciousness and communication but can be sustained as meaningful only by a constant differentiation from the meanings currently possessed, differences that seem to come from "out there." As Luhmann puts it, "the distinction medium/form serves as a frame without outside, as an internal frame that includes, via reentry, its own outside."[31] Meaning is the blind spot generated by the observation of any particular meaning: "Again, the medium is inside and outside, but the attention of the system has limitations and observes only forms."[32]

These formulations resonate with important narrative and media theories regarding the *observation (or not) of the medium*, the oscillation of its status as all or nothing in the observation of communication. Latour will praise the postmoderns for their insistence on observing the modes of mediation that bind subjects and objects, just because the moderns are notorious for dismissing them. Jay Bolter and Richard Grusin's *Remediation: Understanding New Media* begins by sketching some related paradoxes in attitudes toward media through a distinction between *immediacy* and *hypermediacy*. Immediacy is the presentation, through its representation by a medium, of an absent object in such a way that the mediation obliterates its own medium; hypermediacy is the presentation of the medium as an object in its own right prior to its function of representing an absent object. Bolter and Grusin note "a double logic of *remediation*. Our culture wants both to multiply its media and to erase all traces of mediation: ideally, it wants to erase its media in the very act of multiplying them."[33] Much as Latour notes how the absconding of the quasi-objects produced by modern technoscientific processes only multiplies them, Bolter and Grusin note how "the logic of immediacy dictates that the medium itself should disappear and leave us in the presence of the thing represented," while at the same time, "immediacy depends on hypermediacy."[34]

F. K. Stanzel's *Theory of Narrative* provides a narratological version of the same media dynamic. Stanzel advances *mediacy (Mittelbarkeit)* as the defining characteristic of narrative discourse. In the literary medium, it interposes a narrator—"a mediator or 'transmitter'"—between the author, the story, and the reader.[35] As we will discuss at greater length in chapter 3, at this level of textual description, free indirect discourse—what Stanzel terms the "figural narrative situation"—is paradoxically the most immediate form of narrative mediation, as well as the most hypermediate: "in the *figural narrative situation*, the mediating narrator is replaced by a reflector: a character in the novel who thinks, feels and perceives, but does not speak to the reader like a narrator. . . . Thus the distinguishing characteristic of the figural narrative situation is that the illusion of immediacy is superimposed over mediacy."[36]

Or again, the figural narrative situation induces the (dis)appearance of the external narrative medium (the third-person or authorial narrator) by covering it over with the forms of a character's point of view: its internal thoughts and perceptions. Stanzel sees clearly that the notion of point of view "is used in two contexts which are distinct in narrative theory: to narrate, that is to say, to transmit something in words; and to experience, to perceive, to know as a character what is happening in the fictional space."[37] In his system, however, point of view remains a *narrative* function of the text, where its "two functions . . . can overlap. This happens especially frequently where authorial and figural elements . . . appear in close association. In this case the perception of the represented reality takes place from the standpoint of a figural medium, but the voice of an authorial narrator can still be heard in the transmission of the figural perception."[38]

For instance, as described in Stanzel's terminology, figural narrative mediation through a reflector character sets the narrative tone of Octavia Butler's Xenogenesis trilogy at the beginning of the first volume, *Dawn*. Having awakened in a strange room, the character Lilith Iyapo gradually remembers other times she had previously awakened in the same unknown space. *Dawn*'s authorial (third-person, or extradiegetic) narrator exploits the figural situation to shift itself out of view as it "transmits" Lilith's recollections of the voices of her unseen interlocutors:

> She imagined herself to be in a large box, like a rat in a cage. Perhaps people stood above her looking down through one-way glass or through some video arrangement.
>
> Why?
>
> There was no answer. She had asked her captors when they began, finally, to talk to her. They had refused to tell her. They had asked her questions. Simple ones at first.
>
> *How old was she?*
>
> Twenty-six, she thought silently. Was she still only twenty-six? How long had they held her captive? They would not say.
>
> *Had she been married?*
>
> Yes, but he was gone, long gone, beyond their reach, beyond their prison.
>
> *Had she had children?*
>
> Oh god. One child, long gone with his father. One son. Gone. If there were an underworld, what a crowded place it must be now.[39]

Most of the perceiving narrated in this passage is channeled through Lilith's self-perception as she pseudo-narrates without narrating, restaging scenes of dialogue through memory, in a flow of signifying forms momentarily allowing the authorial narrative medium to abscond. As conceptualized by Stanzel's figural mediator, however, narration and perception—even as attributed to two distinct sources—merge into one.

3. NARRATION AND FOCALIZATION

One of Genette's most far-reaching proposals in *Narrative Discourse* is on the concept of *focalization*. Its purpose is to resolve just the sort of analytical ambiguity that occupies Stanzel's presentation of the figural narrative situation, "a regrettable confusion between what I call here *mood* and *voice*, a confusion between the question *who is the character whose point of view orients the narrative perspective?* and the very different question *who is the narra-tor?*—or, more simply, the question *who sees?* and the question *who speaks?*"[40] For Genette, focalization is set apart from narration as a modal measure-ment of the extent to which diegetic characters are allowed to constrict the observation of the storyworld. Genette's scheme measures narrative infor-mation about seeing and knowing on a scale placing the omniscient narrator (that sees and knows all) at zero. For Genette, focalization "is essentially . . . a *restriction*"[41] of the narrator's cognitive prerogatives. What Stanzel terms the figural mediator or reflector-character is for Genette an agent of inter-nal focalization binding narrative information to a more or less limited perspective.

Mieke Bal's vigorous development of the theory of focalization lifts it away from Genette's discussion of narrative mood and grants it discursive autonomy as a principle nearly coeval with narration. "Whenever events are presented, they are always presented from within a certain 'vision.' "[42] Compare Maturana and Varela's well-known neocybernetic axiom on the structural determination and hence material specificity of any possible ob-serving system: "*everything said is said by someone*" (italics in original).[43] Had this been offered as a narratological remark, in Genette's scheme it would apply to matters of voice—in Bal's, to the layer of the text. In parallel with Maturana and Varela, Bal's statement above could be rephrased: *everything seen is seen by someone* (narrator, character, or actor) *from somewhere* (within or without the storyworld). Accordingly, with regard to her three-layer model, for Bal focalization is an aspect of the story signified by the text. It is one of the primary devices whereby the fabula is manipulated in the story layer, precisely by having the events of the fabula coordinated with specific agencies of perception and their particular angles and perspectives:

> Focalization is the relationship between the "vision," the agent that sees, and that which is seen. This relationship is a component of the story part, of the content of the narrative text: A says that B sees what C is doing . . . . The speech act of narrating is still different from the vision, the memories, the sense percep-tions, thoughts, that are being told. Nor can that vision be conflated with the events they focus, orient, interpret. Consequently, focalization belongs in the story, the layer between the linguistic text and the fabula.[44]

Compare the well-ordered cogency of Bal's treatment of focalization within her three-layer model to the difficulties encountered when a group of veteran narrative theorists approach the same issue through the two-component discourse–story model. In *New Perspectives on Narrative Perspective*, a 2001 volume drawn from a 1995 international conference held in the Netherlands, the initial section on "Concepts" wages a vigorous debate over whether narrators can also be focalizors. In his contribution, Gerald Prince backs Seymour Chatman's position that, in Prince's words, "the narrator . . . is *never* a focalizer . . . simply because, qua narrator and regardless of his or her (homo- or hetero-, intra, or extra-) diegetic status and narrational stance, s/he is never part of the *diegesis* she presents. Qua narrator, s/he is an element of *discourse* and not *story* (of the narrating and not the narrated) whereas focalization is an element of the latter."[45]

In his counter-argument, James Phelan does not object to Prince's assignment of focalization to the storyworld (Bal's fabula) and the resulting conflation of the diegetic knower with the diegesis known. Rather, and not without good reason, Phelan remarks how "the Prince/Chatman view of narrators reveals some problems in the underlying principle upon which it is based: the explanatory power of the story/discourse distinction. . . . The story/discourse distinction is a heuristic construct rather than a natural law."[46] Phelan is right to complain that Prince is overreading the discourse–story dyad as an ontological statement rather than as the epistemological construction it clearly is. But Phelan does not appear to grasp the constructivism implicit in his criticism and for no good reason regresses instead to a humanist stance against conceiving the narrator as nonperceiving automaton: "A human narrator, I submit, cannot report a coherent sequence of events without also revealing his or her perception of those events. . . . Indeed, the distinction between perceiving and reporting . . . is ultimately impossible to maintain unless we reduce all narrators to reporting machines."[47]

Out of this seemingly innocuous technical debate Phelan conjures up a cybernetic specter to defend "human" narrators against the insentient "reporting machines" created by the theoretical separation of "perceiving and reporting"—focalization from narration. But one needn't be a dedicated posthumanist to insist on the widely adopted position that Bal properly reformulates at the outset of her *Narratology*: in the literary medium, the narrator can only be a "linguistic subject, a function and not a person."[48] Equally seriously, Phelan's curious histrionics obscure the part of the argument that Prince does get right. Restating the well-vetted Genettian principle, Prince argues that "maintaining a strict distinction between 'who speaks' and 'who perceives,' narrator and focalizer . . . makes it clear that focalization is autonomous with regard to medium of presentation."[49]In systems-theoretical terms, this formulation separating narrative presenting from narrated perceiving conforms to the autonomous operations of social and psychic systems respectively, and the corollary distinction between communication and perception as incommensurable products of discrete system functions. So Prince is right to insist on that separation, but not to draw from it the need for a theoretical restriction cordoning narrators off from focalizors. One can keep the systems references distinct while still entertaining (in Luhmann's technical term) their *interpenetration*, the circumstance that social and psychic systems are tightly coupled and indeed coevolved metabiotic phenomena. Prince maintains and develops Genette's seminal insight in distinguishing narration and focalization, but hardens it inappropriately and forces it onto the wrong template when he claims for this "restrictive definition . . . the advantage of placing focalization squarely in the domain of the narrated or story."[50]

By story here Prince denotes of course Bal's fabula, and she is clearly correct in her counterstatement that focalizations should not "be conflated with the events they focus, orient, interpret." Her tripartite scheme provides a better categorization: "What has been said about any narrative holds for images as well: the concept of focalization refers to the *story* represented and the concept of narrator to its representation, by acting as the steering perspective on the events (or *fabula*)"[51] (my italics). In other words, both narration and focalization are *discursive* components, the former given by the text and the latter as "the direct content of the linguistic signifiers."[52] The text signifies the story directly and relatively immediately; in contrast, a fabula emerges only through indirect mediations. To construct a fabula, the discourse mutually constituted by text and story—narration and focalization—must then be submitted to the selective cognitive operations of analytical interpretation. Discourse—text and story—is the relatively stable term, and any given discourse can yield any number of plausible fabulae: "The events selected can be related to one another in a variety of ways. For this reason, one should not refer to *the* structure of a fabula, but to *a* structure."[53]

Finally, Phelan is right to reject Prince's restriction of focalizing functions from narrators, but not to draw from that discursive coproduction a nostalgic reification of a unitary human subject sentimentally identified with linguistic textual functions. Maintaining "the distinction between perceiving and reporting" will in no way "reduce all narrators to reporting machines." Rather, maintaining that distinction enables the productive alignment of narratological with systems-theoretical distinctions, and reinforces the posthumanist understanding that "persons" are always already social as well as psychic constructions constituted in and by an assemblage of autonomous systems embedding them in complex nonhuman environments.

Let us go one step farther. Neocybernetics makes an important intrasystemic distinction between the functions of *operation* and *observation*: systems must operate to function at all, but those operations are steered to a greater or lesser extent by the particular operation of observation. Focalization complexifies the category of narration by distinguishing a separate, embedded layer of narrative function. The production of the text through its narrating instance corresponds to the self-productive operation of autopoietic systems generally (social systems in particular), as focalization corresponds to the specific operation of perception (in psychic systems in particular). Although focalizations must be narrated (rendered at the layer of the story by the signifiers of the narrator's text), what *is* narrated is a result of what is or is not focalized. Observations must be produced by operations that are guided by those same, delimited observations.

Bal's theory of narrative layers	text: narration	story: focalization	fabula: events, objects, actors, locations, things
narrative processes viewed intra-systematically	narrative operation	narrative observation	(the storyworld)
Luhmann's theory of autonomous systems and their environments	the social system: communication	the psychic system: consciousness—perception, affect, thought	the environment at large
Bal's idiomatic restatement:	"A says . . .	"that B sees . . .	"what C is doing"

Table 1.2. Posthumanist Narratology

In both the narrative and systemic instances, one discerns interpenetration, a double feedback loop in which different systems are recursively coupled together in their inputs and outputs. Or again, the points of view presented by a narration are storybound in a way that the narrative text is not. But because focalizations—the perceptual possibilities and selections actually communicated by the narrator and the characters narrated—constrain the text that creates them, focalizations tell a story *about* the text, running alongside the story told *by* the text: "The point of view from which the elements of the fabula are being presented is often of decisive importance for the meaning the reader will assign to the fabula."[54] One can call the implications encoded into the story layer, along with the ever-present potential for embedding across textual levels, an instance of narrative self-reference. By pressing on the discursive selections both text and story—communicative presentations and omissions, and perceptual openings and closures—the unity of the narration–focalization distinction expands the observability of narrating systems while helping to make sense of narrative blind spots.

4. THE ENVIRONMENT OF THE FABULA

Let us return to the narratological thread of the discourse–story distinction, with particular emphasis now on the concept of fabula (Genette et al's story). Marie-Laure Ryan has recently made a cognitivist intervention in this discussion. With specific reference to Abbott's presentation of the discourse–story distinction from the *Cambridge Introduction to Narrative* cited above, Ryan objects that

> the two components of narrative play asymmetrical roles, since discourse is defined in terms of its ability to represent that which constitutes story. This means that only story can be defined in autonomous terms. As we have seen, Abbott regards stories as sequences of events, but this characterization cursorily equates stories with events, when events are in fact the raw materials out of which stories are made. So what is story if it is not a type of thing found in the world, as existents and events are, nor a textual representation of this type of thing (as discourse is)?[55]

What is again at issue is the ontological status of events as inducted into the milieu of narrative stories (fabulae), where it is generally understood (unless otherwise stipulated) that the events in question, however much they may resemble or take their cue from real events, are imaginary, or at least re-imagined. Ryan continues:

> Story, like narrative discourse, is a representation, but unlike discourse it is not a representation encoded in material signs. Story is a mental image, a cognitive construct that concerns certain types of entities and relations between these entities. Narrative may be a combination of story and discourse, but it is its ability to evoke stories to the mind that distinguishes narrative discourse from other text types.[56]

Paul Cobley has affirmed that "narratology generally embraces the 'constructionist' perspective as its guiding principle," crediting "the thoroughly social nature of the *construction* of meaning."[57] But social constructionism without further methodological armature is at best a loose conceptual program. In his other narratological work Cobley works with the mature research program of semiotics. For her part, in the formulation above Ryan explicitly gestures toward cognitive science to argue for the "autonomy" of narrative stories (fabulae) as internally constructed mental events distinct from outer-worldly historical events.[58] This could be viewed as a somewhat more rigorous version of the narratological truism Abbott states in his introductory text, that "we never see a story directly, but instead always pick it up *through* the narrative discourse. The story is always mediated . . . what we call the story is really something that we construct."[59]

In any number of ways, then, narratology is well en route to—but, as a rule, has yet to arrive at—the self-consistent epistemological constructivism of systems theory. What holds it back, it seems to me, is a common stock of humanist tropes ("the life of the text," "universal masterplots," etc.), and also, the maintenance of a undertheorized representationalist orientation that elides the systemic nature of the processes involved and takes the operational contingencies of cognitive and communicative productions for granted. The productions of autopoietic observing systems involved in any narrative phenomenon are not representations or mere decodings of received transmissions—they are system-specific reconstructions.

An autopoietic system, according to Luhmann, is one "that produces and reproduces through the system everything that functions for the system as a unit."[60] As the ongoing product of their own production, such systems are self-referential. The operational closure of self-referential autopoietic systems means that they do not operate beyond their own boundaries: their closure is the condition of their possibility. Social and psychic systems construct their own meanings out of their own internal elements. "A systems-theoretical approach emphasizes the *emergence of communication* itself. Nothing is transferred"[61]; everything is (re)constructed on the fly. Communication obviously occurs, but it is self-constituting and self-perpetuating. "*Only communication can communicate*"[62] (italics in original). Psyches reside in the *environment* of social systems as separate but coupled systems to which communications come from the social systems in *their* environments. Consciousness as such does not impinge directly upon its social environment, and the communications (re)produced by social systems do not impinge directly upon the minds in their environment. Rather, they impinge, in the forms properly constituted by the autopoiesis of psychic systems, as internally constructed perceptions and cognitions. Whatever consciousnesses of their environments psychic systems achieve are always emergent and selective performances, but are nevertheless performances that mediating structures such as narratives can potentially bring into resonant redundancy.

> But all the story of the night told over
> And all their minds transfigured so together,
> More witnesseth than fancy's images
> And grows to something of great constancy.[63]

An important axiom of Luhmann's theory is that a system must reduce its own complexity relative to that of its environment. The environment of any system is always more complex than that system—not the least because its environment will contain many other systems. Its internal construction (cognition) of its environment will inevitably be incomplete, but that enforced selectivity is a positive outcome in that it guarantees a reduction of complexity. From this angle, Mieke Bal's methods of fabula construction by rigorous analytical selections are also unusually resonant with systems-theoretical constructivism. In her approach (synthesizing a large body of prior research), a fabula is "the 'deep structure' of the narrative text."[64] It is an enforced abstraction from the relative complexity of discourses typically presented by vast profusions of textual signifiers. The object is to winnow the multiplicity of the text down as far as possible without losing the functional specificities of the narrative at hand, and this can always be done in any number of ways. Bal's concept of fabula is not simply the story returned to chronological sequence (Genette's anachronic discourse returned to the chronology of story). Rather, it is an analytical construction resulting from a deliberate condensation—a *reduction of the complexity* of the discourse (text and story) to a set of propositions about its functional fable.

Perhaps paradoxically, alongside the system of communication through which the text at hand is negotiated, one may construct refined cognitions of the text's worldly environment—may grasp in more scientific fashion some of that wider world's systematicities. "Everything that can be said about the structure of fabulas also bears to some degree on extra-literary facts. . . . The theory of [fabula] elements, even more generally than that of [story] aspects, makes describable a segment of reality that is broader than that of narrative texts only."[65] Its possible fabulae are the transmedial components of any narrative instance. Ryan appears to concur in her own terms when she remarks that " 'story' is a cognitive structure that transcends media, disciplines, and historical as well as cultural boundaries."[66] We would just want to stipulate that the transcendence of the narrative by its fabulae has to do with the operational (not ontological) autonomy of the cognitive system that constructs it. That said, it may be that narratives connect to worldly systems not in their putative representational verisimilitude—especially if the narrative communication at hand is fantastic, speculative, or science-fictional—but in the ways that, at their deepest levels of abstraction, they allow the construction of functional homologies to real processes of life, mind, and society.

Extraterrestrial Encounters

The following three mainstream science fictions from the late nineteenth to the later twentieth century recount different narratives of confrontation between human and alien societies. The ideology of these tales of posthuman encounter, however, is largely one of untranscended anthropocentrism. The variously imagined intercourse with alien beings leads not to the overcoming but the reaffirmation of humanist idealisms. This occurs both when species metamorphosis is *not* imagined, as in *The War of the Worlds* and *Contact*, and when it is, as in *Childhood's End*'s climactic psychic merger of the children of Earth into the Overmind. But we will apply our posthumanist narratology to read some of the ways their interplays of narration, focalization, and event construction elicit and encode other messages about psychic and social systemic operations. Through the alienation of focalization, each of these narratives postulates a posthuman perspective—an ultimate knowledge on the part of projected aliens beyond the closure of the human cognitive system. The cosmic perspective thereby achieved allegorizes the normal, posthumanist situation in which communication always already happens.

1. *THE WAR OF THE WORLDS*

Let us consider the opening of H. G. Wells's *The War of the Worlds* (1898) as a scene of focalization: "This world was being watched keenly and closely by intelligences greater than man's . . . perhaps almost as narrowly as a man with a microscope might scrutinize the transient creatures that swarm and multiply in a drop of water. . . . Intellects vast and cool and unsympathetic, regarded this earth with envious eyes."[67] In this account of the now-vanquished Martians, they are focalized as focalizors. This doubled gaze is the reconstruction of the novel's unnamed character-bound narrator, who narrates at a temporal distance of six years from the events he witnessed. In these opening lines, he narrates in impersonation of an authorial narrator peering down on the Martians as they must have been peering down on an oblivious objectified humanity. In this reconstruction, the Martians momentarily occupy the epistemological position of scientific observers, figures licensed under positive modern regimes of objectivity to control the gaze and take it as far as their instruments can go. Yet in the same gesture the narration takes control of and renders them as objects mastered by the narrator's virtual gaze. They are not scientifically proper impersonal observers but "envious" gazers from a dying planet coveting the natural resources of a foreign world, doing military surveillance for an ultimately thwarted invasion.[68]

A centerpiece of the novel is the long episode during which the narrator turns the tables of focalization on the Martians again by occupying a visual vantage from which to observe *them* at work. In the chapter "What We Saw from the Ruined House," the landing of a fifth cylinder of Martians demolishes the autodiegetic narrator's current shelter, but propels him to a protected point overlooking their war preparations: "We hung now on the very edge of the great circular pit the Martians were engaged in making.

The heavy beating sound was evidently just behind us, and ever and again a bright green vapor drove up like a veil across our peephole."[69] The novel as a whole presents itself as an occasion for the narrator to refocalize and then verbalize memories generated by having directly observed the mysteries of the alien beings, so that, after the fact, social communication can construct a victorious interpretation of events.[70]

In *The War of the Worlds* there is sentience on either side of the conflict, but no will toward any peaceable interaction, any negotiations that could broker an extraplanetary expansion of human society. Much like modern sciences' tendency to monopolize epistemological authority, the narrator's monopoly on focalization demotes the Martians to the status of objects, and since they never discourse with the narrator, the Martians remain the objects rather than coagents of his reality constructions. Specular objects indeed, upon which he projects tendentious evolutionist imageries, but not, however, without a curious proto-Gaian twist in which the microbial inhabitants of the Earth rise to the defense of their terrestrial brethren:

> Stark and silent and laid in a row, were the Martians—*dead!*—slain by the putrefactive and disease bacteria against which their systems were unprepared. . . . These germs of disease have taken toll of humanity since the beginning of things—taken toll of our prehuman ancestors since life began here. But by virtue of this natural selection of our kind we have developed resisting power. . . . But there are no bacteria in Mars, and directly these invaders arrived, directly they drank and fed, our microscopic allies began to work their overthrow.[71]

Like many adventurers into dark continents before them, dispatched by nonadaptation to an alien environment, the Martians die of unforeseen pathogens in the bush of a hostile would-be colony. From the beginning of this narrative, "ethnic" (barely-veiled imperial) hostility short-circuits symbiosis and the potential for sociality.

2. CHILDHOOD'S END

Let us contrast *War of the Worlds* to a novel written two world wars later: Arthur C. Clarke's *Childhood's End*. Here the will of separate worlds to converse is palpable. The Overlords arrive not to conquer but to deliver a dark message. In their ships they seem to hover over the entire globe, but they will not show themselves. From their own high enclosures, they will only converse, and with only one human being: "The Secretary General of the

United Nations stood motionless by the great window, staring down at the crawling traffic on 43rd Street. He sometimes wondered if it was a good thing for any man to work at such an altitude above his fellow humans."[72] Brought forward as the proxy human focalizor, the character Stormgren is the medium for the international authority of the spoken conversation between humanity and the Overlords.

Above Stormgren, Clarke places an authorial narrator capable of rising to a level of observation higher than that of any character, alien or human. This observing frame is keyed to the extraterritoriality of the "international observers" instituted at the United Nations not long before the novel was first written in 1953. This narrator's cosmic internationalism drives its strong thematization of scientific observation and its multiple delegations of focalization. Most consummately, the young researcher Jan Rodricks stows away in an Overlord lightspeed cruiser for a six-month/eighty-year roundtrip to the Overlords' planet, and earns a series of increasingly higher visions, witness of a power higher than the Overlords, and finally, upon returning, witness of the last human being left behind at the cataclysm that apotheosizes the human spirit and consumes the Earth. But the final point of view of the narrative is given to the alien being who outlives the era of the humanity's terminal evolution into the mystic substance of the Overmind. From his starcruiser at the edge of the solar system, "for a long time Karellen stared back across the swiftly widening gulf, while many memories raced through his vast and labyrinthine mind. In silent farewell, he saluted the men he had known."[73]

In the last analysis this authorial voice operates from the level of the glimpsed but essentially unrevealed Overmind; whatever conversation takes place between humanity and the Overlords is ultimately moot. The communication that matters is between the Over*mind* and the human spirit, and that channel is already utterly mystified, rendered entirely in transcendental media, replete with Ouija boards and telepathic circuits. In a kind of Anglo-American United Nations wish fulfillment, the social communications needed to heal the divisions of a Cold-Warring and decolonizing world descend from the cultural heights of the galactic first world to the underdeveloped backwater of Earth on wings on mystic good will alone. The structure of authorial narration supports the social mystification divulged in this narrative: by reading characters' minds, authorial-narrative mediations render psychical what are in reality social operations. Operations of communication are attributed to organs of consciousness that are in closed-operational fact unable to support them.

The fictive omniscience of the authorial narrator—that can, in this text, make a report on the thoughts of an alien Overlord waving humanity good-bye—renders fictionally literal the notions of mind reading and telepathic communication.[74] Still, as in *The War of the Worlds*, the reversal of observation, the projecting of focalization beyond the world to get a look back *at* the world, is an important gesture. The alienation of focalization is an important service of science fiction narrative as an active mode of social self-reference. It implies the alienation of narration as well, the implication being that a narrator occupying an observing frame beyond alien story-worlds that are themselves out of this world is also beyond the limits of human comprehension, and therefore is "authoritative" about ultimate things.

3. CONTACT

We see the same narrative gambit shot through the text of Carl Sagan's *Contact* of 1985. That novel packages its narrative in multiple epigraphs, but also, for the first four chapters, sets forth an additional framing device in blocks of indented italics. In the first block, the cosmic authorial narrator focalizes an artificial planet, a galactic communications satellite orbiting in the constellation Vega, and renders some further knowledge about its internal operations: *"The polyhedral world had been performing its enigmatic function for eons. It was very patient. It could afford to wait forever."*[75] Then, having received notice of technological intelligence on Earth—electromagnetic signals of the maiden TV broadcast of Hitler opening the 1936 Olympic Games—the preprogrammed communications planet sends a message back to Earth containing the schematics for a mysterious machine. For three more chapters, the cosmic narrator tracks those signals as they approach the Earth at the speed of light, to be presently intercepted by SETI researcher Ellen Arroway.

As a girl, Ellie was already adept in constructing scenarios of reverse focalization: in her fantasies about a solar system teeming with life, she "would imagine a young [Venusian], glancing up at a bright blue point of light in *its* sky, standing on tiptoe and wondering about the inhabitants of Earth."[76] SETI is a wonderful real-world crystallization of the ironies of observational regimes. Here with a vengeance is the romance of meaning—the cosmic desire for a meaningful message from an exotic extraterrestrial beyond. And the use of radio astronomy not just to observe the universe in a detached manner that implicitly rises above the universe under observation, but to *listen for a signal denoting alien intelligence*, is indeed to move in a radical way away from a colonial and toward a collaborative model of science as a social activity. Thus it is a letdown when Ellie, grown up now and a courageous and embattled female scientist, projects upon her would-be extraterrestrial conversational partners a superiority of mind that licenses them to colonize *us*: "But surely, she thought, they would know better than we what modulation frequencies were acceptable. . . . They would know how backward we are, and would have pity."[77] Later, looking down from an airplane window, Ellie again places herself, as proxy Earthling, in the role of an object to be evaluated, and projects her vision to the level of the alien subject performing that evaluation: she wonders "what impression the Earth would make on an extraterrestrial observer. . . . Would they be able to tell in one swift glance precisely what stage we were in some great cosmic evolutionary sequence in the development of intelligent beings?"[78]

By the end of the narrative, of course, humanity does pass the test. In the novel (as opposed to the 1997 film version) a *crew* of cosmonauts take the ride on the machine built from the alien blueprints, and so earn their opportunities, like Jan Rodricks in *Childhood's End*, to focalize the further stretches of the galaxy. This experience brings Ellie, the inveterate reverse focalizor (and in this passage, reflector-character), face to face with her mirror image as a radio astronomer. From the outskirts of the star Vega, the crew observes out the window of their vehicle:

> Xi, whose eyesight was acute, was staring up. . . . Ellie followed his gaze. . . . She looked through the long lens. It was some vast irregular polyhedron, each of its faces covered with . . . a kind of circle? Disk? Dish? Bowl?
> "Here, Qiaomu, look through here. Tell us what you see."
> "Yes, I see. Your counterparts . . . radio telescopes. Thousands of them, pointing in many directions. It is not a world. It is only a device."[79]

By this device—a cosmic "mirror site" reflecting the protagonist's desire to see her own reflection in the stars—at its climax as an adventure narrative, the novel raises Ellie's character to within one frame of the narrator, from which vantage she can herself momentarily focalize the same object presented at the beginning of the text. That polyhedral planet is a mirror not only of the main character Ellie but also of the authorial narrator: both the narrator and the radio planet are receiver-focalizors and transmitters of intelligent signals. The planet is also a marker for the galactic civilization (occupants of the narrative's unmarked space) that constructed it and placed it on sentinel duty. When ultimately, on the virtual beach, a liminal zone for worlds in contact, Ellie actually does have a conversation with the intelligence behind the Vega satellite (in the overdetermined form of her dead father), what she learns is:

> "We collect information. I know you think nobody has anything to learn from you because you're technologically so backward. But there are other merits to a civilization."
> "What merits?"
> "Oh, music. Lovingkindness. (I like that word.) Dreams . . ."[80]

Here is a further evolution of the benevolent alien beloved of the Western liberal imagination, a secularized ideal loving father whose gifts outweigh his demands. Again, as in *Childhood's End*, the upshot of the science-fictional upshift to a scene of galactic conversation between separate worlds is the discovery that, like Clarke's Overlords, Sagan's message-senders are "just caretakers"—just forms appearing in a medium of ultimate meaning that remains unseen, unattainable, untunable into. There is another order of being beyond the galaxy, another Message beyond the message, another Overmind over any other mind. Thus the matter of alien sentience is thrown back down the black hole of an epistemological mise en abyme, and we end up, like Ellie, back where we started from, with only human memories by which to nurse along a faith in the higher intelligence of the universe.

Each of these science-fiction narratives uses scenarios of cosmic focalization to stage extraterrestrial allegories of social communication. But, observing these narratives in systems-theoretical terms, one must ask: could the social autopoiesis that has arisen on Earth ever really process such radical alienations of its environment as those in these alien scenarios? In "The World Society as a Social System," Luhmann takes a kind of global view:

> Society is an exceptional case. It is the encompassing social system which includes all communications, reproduces all communications and constitutes meaningful horizons for further communications. Society makes communications between other social systems possible. Society itself, however, cannot communicate. Since it includes all communication, it excludes external communication. It has no external referent for communicative acts, and looking for partners would simply enlarge the social system. . . .[81]

In Ellie's constructions of desire and the SETI gesture generally, beyond the society of global communication (for which human bodies and minds are the environment) is posited another society, a society beyond the world that could itself open itself to an interchange with another global system beyond *its* own systemic borders.[82]

> This, of course, does not mean that society exists without relations to an environment or without perceptions of environmental states and events; but input and output are not carried by communicative processes. The system is closed with respect to the meaningful content of communicative acts. The content can be actualized only by circulation within the system. At the same time, but at another level of reality, the system uses body and mind of human beings for interaction with its environment.[83]

The narrative quest for alien communication, "looking for partners," is an allegory of posthuman systemic merger. Meanwhile, right here, in our bodies and our thoughts, we *are* being watched and manipulated. Without our awareness, social systems virtually invade our bodies and minds. They focalize us, producing inhuman narrations that contain us. We are the environmental references maintaining the self-reference of social systems binding individuals by the mediatic threads of communicative events. We insist that when communication occurs, minds meet. The fact is that there is no contact. Rather, minds mutually self-construct at a distance with the environmental support of the media of communications in which they are immersed. No matter how powerful our eyes on the sky get to be, to be a part of our society, the universe will have to come to us.

TWO

Nonmodern Metamorphosis

> We shall at least be freed from the vain search for the undiscovered and undiscoverable essence of the term species.
>
> — CHARLES DARWIN

> What on earth was he—man or animal?
>
> — H. G. WELLS

Narrative, Networks, and Systems

The theory of evolution marks a recent moment in the long history of narratives of bodily metamorphosis, a modern moment when scientific discourse presented new and persuasive explanations for divergences in the forms of living beings. Darwinisms of many stripes replaced more traditional ways of accounting for magical or uncanny changes of species. Technological developments since then, such as television and space travel—magical indeed by traditional standards—added machines to the ranks of "evolutionary" entities. The discourse of cybernetics emerged at mid-twentieth century to explore the increasingly complex interfaces of technological and biological systems. Second-order systems theories mark a neocybernetics focused on the formal conditions and structural couplings of system–environment multiplicities.

Working from comparable positions put forward by philosopher of science Michel Serres, sociologist of science and technology Bruno Latour's

works are also inspired by neocybernetics. Latour's deeply embedded connections to neocybernetic discourses inform his important polemics against
philosophies that divide beings up and "purify" nature and society one from
the other. His arguments both complement and qualify more explicit systems theories that stress equally strongly the bounded autonomy of autopoietic or self-producing systems. The discourses of hybridity in networks
and the discourses of self-referential closure in systems stem from the same
set of revisions to classical cybernetic sources, and they can work together
in a more generously conceived neocybernetics. As Niklas Luhmann often
stresses in discussing the interpenetration of different and autonomous systems, each operates to compensate for the other's blind spots. Stories of
bodily metamorphosis, especially but not exclusively those contemporaneous with cybernetics, may often be seen as fantastic allegories of these mundane but generally unconscious and uncanny interconnections.

Latour's later sociological methods coalesced around a comparable allegory of repressed connections in a conceptual figure derived from Michel
Serres.[1] While initially focused on "actor-network" theories of fact construction in technoscientific collectives, Latour proceeded to construct and
observe the intermingled operations of natural and cultural formations in
scientific research and technological projects by following the circulation of
quasi-objects.[2] The quasi-object names the objecthood of subjects (such as
human persons) and the subjecthood of objects (such as machines and nonhuman organisms). And the stories Latour tells about quasi-objects present
a range of transformative interactions that rhyme conceptually with classical
and modern narrative fictions of metamorphic changes. The metamorphic
characters in such stories are *fictive* quasi-objects—manifest hybrids, usually
of human and nonhuman components, brought about by various magical or
daemonic technologies, beginning with the technologies of language and
narration per se. This already suggests something anachronistic about
quasi-objects—they capture or rejoin a perennial idiom of narrative
mythopoesis.

The Island of Dr. Moreau is a modern story in this perennial metamorphic
vein—that is, modern in Latour's sense. Even while it reveals in the Beast
People the hybridity of technoscientific quasi-objects, its eponymous main
character oppresses them and its narrator represses their nonmodern significance. This metamorphic narrative virtually produces a template for the
residual anthropocentrism of mainstream science fictions ostensibly devoted to repudiating the essentiality of the human. *Dr. Moreau* is also inter-
esting as an allegory of modernity because its metamorphs are "evolved"

rather than debased—anthropomorphized animals rather than animalized humans. A reluctant witness of Dr. Moreau's covert operations, the uninvited castaway Prendick at first mistakes the ontological status of Moreau's Beast People for the latter. The climactic revelation that animals *are* humanizable, and by humans themselves, despite whatever cruelty is involved, strikes him as confirming a darkly metaphysical interpretation of the theory of evolution: that without a stable demarcation between the human and the animal, the human will not stay human but "lapse" into bestiality, and that this boundary has already been breached. The final words of his written narration moralize this melancholy apocalypse: "There is, though I do not know how there is or why there is, a sense of infinite peace and protection in the glittering hosts of heaven. There it must be, I think, in the vast and eternal laws of matter, and not in the daily cares and sins and troubles of men, that whatever is more than animal within us must find its solace and its hope. I hope, or I could not live. And so, in hope and solitude, my story ends."[3]

Latour's writings suggest instead that, as literary metamorphs with the specific shape of technoscientific quasi-objects, the Beast People have something important to say about how human, natural, and technological systems actually function and interact, something to say about worldly sociality that Prendick cannot see. Of quasi-objects, as we will see later, Latour speaks about *(x)-morphism*—the *x* factor being the play of indifference between *subjects* and *objects* when it comes to the construction of sociotechnological networks such as scientific laboratories, engineering projects, and the human and natural communities that now depend on them. Through the observation of quasi-objects, one recovers not a human–nonhuman standoff but a "variable-ontology world . . . the result of the interdefinition of the actors."[4] Latour's highly mobile concepts describe a neocybernetic vision of the necessary hybridity of symbiotic networks and system–environment couplings, and they describe equally well the daemonic landscapes of metamorphic narratives.

Allegory of Narrative

Like the vicissitudes of persons in love or conflict, the vicissitudes of bodies are cornerstones of narrative fabulae. Body changes may play out as the representation of familiar corporeal experiences—as aging, as the renovations of puberty or pregnancy, or as the result of the mundane violence of other persons, physical forces, living processes, or cultural models: for instance, injury, illness, or body-deforming constrictions of labor or social role. But in mythic and fantastic narratives, bodily metamorphoses take paradoxical turns and play out as impossible or contradictory physical changes.

In "Cybernetics and Ghosts," Italo Calvino refers such fantastic events, all the mayhem of mythic or magical transformations, to a mode of self-reference inherent in the transmission of stories—the storyteller's primal focus on language itself, the narrator's capacity for countless constructions and recombinations among the elements of the media of narration. In the construction and oral delivery of fables and myths, the

> immobile world that surrounded tribal man, strewn with signs of the fleeting correspondences between words and things, came to life in the voice of the story-teller, spun out into the flow of a spoken narrative within which each word acquired new values and transmitted them to the ideas and images they defined. Every animal, every object, every relationship took on beneficial or malign powers that came to be called magical powers but should, rather, have been called narrative powers, potentialities contained in the word, in its ability to link itself to other words on the plane of discourse.[5]

Magical or daemonic events "on the plane of discourse," then, are both the cause and the effect of the capacity of language in social circulation to sound out and link up its own structures. The narrative depiction of fantastic bodily metamorphoses sets into further play the formal possibilities of linguistic and conceptual combinations. Moreover, the narrative drive toward images of bodily transformation tests and contests the boundaries of "identities" and their psychic and social regimes. Transformation stories are a kind of social-systemic program tool for tweaking the cultural hard drive. But what is the "spirit" or "daemon" that calls forth these aberrant figurations? Of their evolution—the transformation of metamorphoses from premodern to posthuman, theological to phenomenological frames of reference—Luhmann might ask: "What is 'Spirit' if not a metaphorical circumlocution for the mystery of communication?"[6]

In a previous book, I wrote that the shapes of literary metamorphoses (that is, post-mythic and post-oral transformation stories) may be read as *allegories of writing*. Literary metamorphoses inscribe fictional bodies with the forms of writing: for instance, the erasure of the prior body through its translation into foreign signifiers. I would extend that reading now by being explicit about the self-reference of narrative texts as complex reflexive elements in the operation of social systems. As oral allegories of speech, written allegories of writing, or cinematic allegories of cinema, the turns of metamorphic stories in any medium are also *narratives of narrative*—self-referential structures, typically stories within stories, that embed the transformations of events within the transformations of characters. That these fantastic narrative events happen at all I take as a systemic response to communicative demands crucial to the self-maintenance of social groups. Metamorphic stories are especially good at the processing of paradox.[7] And, according to second-order cybernetics, paradox is the epistemological non-foundation on which systems stand or fall.

Latour and Metamorphosis

These matters of literary transformation and bodily metamorphosis set up a narrative-systems approach to Bruno Latour's major theoretical statement, *We Have Never Been Modern*, and his literary-experimental sociological study *Aramis, or the Love of Technology*.[8] These works present extended methodological reflections on the sociology and anthropology of technoscience along with a wealth of expository information. But the fictive and nonfictive anecdotes and stories Latour tells in those texts also present transformative actors and interactions—derived from actual scientific and technological practices—that parallel the daemonic agents and subjects of literary and narrative metamorphoses. For Latour, in scientific objects and technological projects, the operations of natural and cultural formations are necessarily intermingled, occupying neither a nature outside of society nor a society outside of nature. To observe the real intermingling of the natural and social, Latour follows the circulation of quasi-objects, as we have noted, entities with indeterminate or multiple references to categories of subject and object. While Latour's quasi-objects are discursive formations constructed from the observation of technoscientific practices, they also join the ranks of literary metamorphs and other fictive actors of transformative fantasies in significantly problematizing ontological distinctions between subjects and objects. The narrative actors of transformation stories share the formal problematics of quasi-objects and -subjects. The latter, however, are also directly referential, or at least something more or other than strictly fictive.[9] On the flipside, Latour's quasi-objects validate the efficacy of fictive metamorphs in symbolically capturing a form—the two-sided (system–environment) form—of systems operations.

But the reality of Latour's quasi-objects also has to do with their local and historical specificity: whatever self-determination the separate elements of an operating worldly network can have can only be a function of particular circumstances and material interdependencies of that network. Thus Latour's proxy in *Aramis*, the sociological mentor Norbert (alluding to Norbert Wiener, the founder of cybernetics), declares that he seeks a " 'refined sociology which applies to a single case, to Aramis and only Aramis. I'm not looking for anything else. A single explanation, for a single, unique case; then we'll trash it.' "[10] Both through the Norbert persona and in his own discursive person, Latour's professed resistance to sociological metalanguage is an important methodological constraint of his network theory. But the narratives energized by this resistance can also be observed through the metalanguage of systems theory, just as the concept of the quasi-object as Latour deploys it can be usefully generalized to the fictive and fantastic constructions of narrative actors. Putting networks and systems together will help us interrogate the real interrelations of narrative and knowledge.

The metamorphic dynamics of Latourian networks occupy three interconnected registers: translation, mediation, and the "redistribution of the human." In his earlier work *Science in Action*, the sense of *translation* is embedded in the sociological analysis of technoscience: the construction of facts by mobilizing the flow of material and mediatic inscriptions from the bench to the textbook. Science in action demands the "translation of interests" by which diverse human and nonhuman constituencies are allied into operational networks. For Latour this "translation model" of science as complexly negotiated material fact construction is set forward against the "diffusion model," a "mentalist" scenario and popular "fairy tale" in which solitary scientific geniuses from Galileo to Einstein come forward as prophets set apart from society, establishing facts of nature solely by turning their visionary powers on key scientific ideas.[11] In this disembodied mode of narrating science, the "scientistics" (the fundamentalists of scientific revelation) purvey an idealist vision of science as a progressive knowledge "diffusing" of its own accord—radiating like a star and moving inexorably forward toward truth, impeded only by the resistance of cultural reactionaries and other bogeypersons, such as Bruno Latour.[12] Latour's quasi-objects inhabit metamorphic narratives intended to disrupt this pervasive mode of scientistic mystification.

In *We Have Never Been Modern*, Latour inflects his demystifications of the diffusion model toward an argument about modernity, and expands the sense of translation from the construction of facts to the construction of quasi-objects, as that is enabled by the "modern Constitution." The agents and subjects of technological networks are nonhumans as well as humans, which then may both be termed, if circumstances warrant, hybrids, quasi-objects, or quasi-subjects. Latour folds self-reference into his discourse at the outset: the science-studies researcher, too, has been transformed by the quasi-object of research: "Hybrids ourselves . . . we have chosen to follow the imbroglios wherever they take us."[13] Latour comes to see this more refined ontological and procedural mode of translation as one of two poles of modern practices:

> The hypothesis of this essay is that the word "modern" designates two sets of entirely different practices which must remain distinct if they are to remain effective, but have recently begun to be confused. The first set of practices, by "translation," creates mixtures between entirely new types of beings, hybrids of nature and culture. The second, by "purification," creates entirely distinct ontological zones: that of human beings on the one hand; that of nonhumans on the other.[14]

Latour constructs modernity as an episteme bifurcated by the terms of an unstated modern Constitution under which foregrounded courses of purification and separation render the operations of hybridity unobservable, allowing quasi-objects and -subjects to proliferate out of sight. The "work of translation or mediation" has made the work of purification "possible: the more we forbid ourselves to conceive of hybrids, the more possible their interbreeding becomes—such is the paradox of the moderns."[15] Maintained by the repressed mediations of absentee progenitors, hybrids proliferate all the more avidly for that lack of chaperoning. Latour thus inscribes modern subjects and objects with paradoxical identity formations, then advances "nonmodernity" as the surpassing of that paradox. In classical psychoanalytic terms, he offers a talking cure for the modern neurosis, an overcoming of modernity's repression of its own technoscientific contingencies.

Mediation, Myth, and Nonmodernity

As we have seen in passages just cited, in the 1990s Latour reframes the transformative dynamics of professional and procedural translations with creative and critical mediations. *Mediation* in this sense operates on the "middle ground" repressed or occulted by the regimes of modern purifications bent on setting apart the human from the nonhuman. Nevertheless, "everything happens in the middle, everything passes between the two, everything happens by way of mediation, translation and networks, but this space does not exist, it has no place. It is the unthinkable, the unconscious of the moderns."[16] Quasi-objects concretize and actualize the formal mediations that hold nature and society together—mediations that were first observed (in the high modernist era of structuralist theory) in the form of *semiotic* phenomena from which material references had been detached. Latour lauds linguistic structuralism and its progeny of postmodern philosophies for taking as their object this middle ground between the modern divisions of nonhuman nature and human society: "The object of all these philosophies is to make discourse not a transparent intermediary that would put the human subject in contact with the natural world, but a mediator independent of nature and society alike."[17]

Linguistic structuralism and poststructuralism showed that semiotic mediators are not docile couriers of meaning but upstart agents with their own semantic agendas. In Latour's later idiom, the *intermediary*—a passive delegate of the diffusion of knowledge, a supposedly reliable messenger—is distinguished from the *mediator*, which always deviates to some degree on its way from source to destination, reworking the given script, the message sent, to translate between and thus connect otherwise uncoupled realms.[18] "The greatness of these philosophies was that they developed, protected from the dual tyranny of referents and speaking subjects, the concepts that give the mediators their dignity—mediators that are no longer simple intermediaries or simple vehicles conveying meaning from Nature to Speakers, or vice versa."[19]

But semiotic mediation is neither as transparent nor as opaque as modern purifiers, semioticians included, would like to think. For Latour, the recently observed significance of semiotic mediation does not discount the real contingencies of the realms being mediated. Postmodern philosophies of language secured their middle ground only by bracketing out the functions of linguistic reference. Their liberation of the "median space between natures and societies so as to accommodate quasi-objects, quasi-subjects" came at a price no longer worth paying: the detachment of linguistic reference from the rest of the world.[20] It is a myth to think, as both the idealists and the materialists seem to do, that there can be elements without mediations. But it is equally inadequate for simplistic deconstructors to think that there can be mediations without elements mediated. Or again, the closure of linguistic reference does not prevent its operational coupling to natural and social systems; without that coupling, language would have nothing to do.

These ambivalences of translational mediation pointed out by Latour—unstable distinctions of social agency between primary and secondary, major and minor, active and passive delegation, the mediator and the intermediary—also typically structure the literature of metamorphic changes, and more generally, the discourse of the daemonic.[21] In classical theological and philosophical mythopoesis, transformative dynamics are a prerogative of the divine parent, but more so, an assertion of the daemonic child. One celebrated avatar of the classical daemonic is the figure of Eros as Plato's Diotima presents it to Socrates in the *Symposium*. Eros comes forward in that dialogue as the proper intermediary carrying messages to and from the human and the divine. The mediator as unreliable messenger is familiar in the classical figure of Hermes, son of Zeus and his sometime herald, who typically ditches his given assignments in favor of amorous escapades, a circumstance brilliantly recaptured for Romantic readers in John Keats's narrative of daemonic metamorphosis, *Lamia*. But in Apuleius's telling of the story of Cupid and Psyche, a long tale embedded within the larger metamorphic farce of *The Golden Ass*, Eros/Cupid is dramatized as Venus's insubordinate son. In more recent literatures, this ambivalence in daemonic mediation often follows the separatisms of Judeo-Christian theology and, preparatory to the course of modern purifications, is parceled out into uncommunicating spheres of the angelic and the demonic.

Latour's intermediaries and mediators, then, reveal their mythopoetic vocation as varieties of the informatic angels and daemons also sighted by Michel Serres and gathered into his multi-volume *Hermes* and his *Angels: A Modern Myth*. The middle ground of the quasi-objects/quasi-subjects figures at once in the daemonic realm of Western mythopoetic anthropomorphosis and as a picture of the material nature of reality construction through communication in an always already-mediated world. The real is what it is, but, as far as we can grasp it, it is also a virtual realm of systemic reconstructions. Thus Latour insists that the networks traced by quasi-objects are "simultaneously real, discursive, and social."[22] They present both media-technological and systems-operational guises that the student of the technosciences must learn to decode and reassemble:

> Such metamorphoses [of quasi-objects] are incomprehensible if only two beings, Nature and Society, have existed from time immemorial, or if the first remains eternal while the second alone is stirred up by history. These metamorphoses become explicable, on the contrary, if we redistribute essence to all the entities that make up this history. But then they stop being simple, more or less faithful, intermediaries. They become mediators—that is, actors endowed with the capacity to translate what they transport, to redefine it, redeploy it, and also to betray it. The serfs have become free citizens once more.[23]

Citizenship in a "republic of things" under a "nonmodern Constitution," Latour concludes, depends on the capacity of the quasi-objects/quasi-subjects to shoulder referential burdens—to bind real connections among natural and cultural agencies.

By the same token, things and persons hold onto existence as far as quasi-objects/quasi-subjects will carry their burden of being: "I call this transcendence that lacks a contrary 'delegation.' The utterance, or the delegation, or the sending of a message or a messenger, makes it possible to remain in presence—that is, to exist."[24] I take this to mean that, while nature does transcend society, and society does transcend nature, neither of these autonomies is purely autonomous. Neither could exist if their differences depended on the negation of the other. Remaining in presence means maintaining the presence of the Other. Natural and social systems both subsist as environments of the other, and the system–environment relation is a two-sided form, a mutual supplementation, a doppelganger and not a dialectical antithesis.

If one allows the extension of sociality beyond human conversations to the communications of other living things—all of which survive by signaling to their own—and to the nonliving things that get swept up and redefined by natural and social systems—then life and its evolution, including the emergence and networking of minds and societies across the living spectrum, is as much a social as a natural phenomenon.[25] "All durability, all solidity, all permanence will have to be paid for by its mediators. It is this exploration of a transcendence without a contrary that makes our world so very unmodern, with all those nuncios, mediators, delegates, fetishes, machines, figurines, instruments, representatives, angels, lieutenants, spokespersons and cherubim."[26]

For Latour, that we can now see (if we wish to look) the intermingled transformativity of natures/cultures marks our status as nonmodern. Our "world ceased to be modern when we replaced all essences with the mediators, delegates and translators that gave them meaning. That is why we do not yet recognize it. It has taken on an ancient aspect, with all those delegates, angels and lieutenants."[27] While this communicational vision is anachronistic, it is neither neopagan nor antimodern—rather, it is *neocybernetic*, a further turn on the conceptual events of the 1960s that Calvino was treating in his coupling of cybernetics and ghosts. When the real and the daemonic are observed emerging and merging in both technological and narrative constructions, classical human persons—the extraenvironmental essences of selves, souls maintained by ideal bodily stabilities—become at once nonmodern and posthumanist, relativized actors performing operational functions and metamorphic transformations within natural/social networks and systems. This is not a demotion of the human but an elevation of the nonhuman into proper discursive representation.

Anthropos and Morphism

Latour calls this ontological condition of medial transformativity "morphism," arriving at that term by deleting from "anthropomorphism" the humanist idealization of *anthropos*. We remain embedded in "the old anthropological matrix,"[28] but the "ancient aspect" of our nonmodern daemonic world is not to be confused with premodern daemonism, which *did* have a contrary—the modern Constitution. In Latour's allegory of real politics, nonmodern morphism arises from the redistribution of being after the reworking of the modern Constitution to convene a new parliament of hybrids. This metaphysical liberalism is Latour's version of the posthuman. While not calling for a technoevolutionary transcendence of the human— why bother when one can have "transcendence without a contrary"?—in this prophecy the human is relativized by its reentry into worldy ensembles with the nonhuman. "Where are we to situate the human? A historical succession of quasi-objects, quasi-subjects, it is impossible to define the human by an essence."[29] Rather, the human demands ongoing nonmodern reassembling—which is to say, in words Latour does not use, that the human is reobserved as a systems phenomenon of autopoietic networks. To maintain its further autopoiesis in the face of its previous autopoiesis (e.g., the rise of modern technoscience), human modernity must be redistributed along the middle ground with the redistributions of the natural and the social:

> If the human does not possess a stable form, it is not formless for all that. If, instead of attaching it to one constitutional pole or the other, we move it closer to the middle, it becomes the mediator and even the intersection of the two. . . . The expression "anthropomorphic" considerably underestimates our humanity. We should be talking about morphism. Morphism is the place where technomorphisms, zoomorphisms, phusimorphisms, ideomorphisms, theomorphisms, sociomorphisms, psychomorphisms, all come together. Their alliance and their exchanges, taken together, are what define the *anthropos*. A weaver of morphisms—isn't that enough of a definition?[30]

To accept this definition is to allow the distinction between the human and the daemonic to lapse: daemonic metamorphosis always was a self-referential projection of the nonessentiality of the human. It is to see that the daemonic situation of medial contingency remains a real allegory of the human, and that this allegory has now been heightened by the proliferation of scientific powers and informatic technologies. "Transcendence without a contrary" means society is maintained only through communication; we communicate only through media, therefore we maintain without surpassing the medial contingencies of the construction of the human—and narrative systems perform this maintenance. "The human is in the delegation itself, in the pass, in the sending, in the continuous exchange of forms," and this status is distributable to everything we touch or that touches us: "Human nature is the set of its delegates and its representatives, its figures and its messengers."[31]

Beast People: The Island of Dr. Moreau

The nonessential or constructed nature of the quasi-object returns us to evolutionary theory. Darwin's *Origin of Species* concludes with a powerful prediction about the intellectual transformations in store, once his explanations of biological form and transformation are accepted: "we shall have to treat species in the same manner as those naturalists treat genera, who admit that genera are artificial combinations made for convenience. This may not be a cheering prospect; but we shall at least be freed from the vain search for the undiscovered and undiscoverable essence of the term species."[32] The quasi-object is one of the later progeny of this liberation from essence. Credence in the origins of species from the environmental selection of genetic variations (and more recently, from the organismic propensity for the viable assemblage of distinct genomes) renders untenable the notions of either a theological or a biological essence of humanity. Evolution unfixes the subject status of the human and the cultural finality of the modern. We are no longer above the beasts, animals no longer merely bestial, and no race or variety of *Homo sapiens* can expect to hold preeminence over another race or species without itself being superseded in turn.

But Wells's Beast People are also the progeny of Darwin's influential and problematic statement of 1872, *The Descent of Man*, in which his cultural conclusions often resemble those of the contemporary religious conservatives who still cast aspersions on his name. *Descent*'s inclusion of the human in the story of evolution pressed Darwin harder toward spatial metaphors of "height" and "depth" to denote greater and lesser evolutionary "perfection." Relative to *Origin*, this rhetoric represents a regressive trend, a residual theologism (or modern Constitutionalism maintaining distinctions between the human and the nonhuman) that leaves its marks on the scientific idealism of Dr. Moreau.

From this angle, *The Island of Dr. Moreau* is a tale about a technoscientific project in "higher" evolution thorough surgical metamorphosis. Dr. Moreau eventually asserts to the character-bound narrator Prendick, "'These creatures you have seen are animals carven and wrought into new shapes.'"[33] The narrative imagines the deliberate transformation of nonhumans, individual animals of various mammalian species, into humanoid beings. While this vivisection plot has typically called forth ethical readings of the human–animal division, what Latour helps us see is that the successes and failures of Moreau's experiments are as much social as natural. That is, Moreau's creations, once set into being, are the result of mediations sustained and relinquished in the networks of communication called forth by the Beast People's own need (in the higher realism of this fable) for emergent systems capable of endowing these newly minted quasi-subjects with functioning social identities. The individualistic Moreau goes only from one surgical subject to the next, abandoning the Beast People to form their own social system based on a common origin.

As descended from another creation-abandoner, Victor Frankenstein, Moreau is no longer a neurotic late adolescent but a degenerate Prometheus, a sociopathic, vivisectionist graybeard—a nonconformist idealist gone bad, like a morbid Thoreau whose *Walden* is not a suburb of Boston but a desert island in the South Pacific. In Latourian terms, Moreau comes forward as a demagogue of the modern Constitution, as his effort to master evolution by purifying the bestial can only manufacture hybrids of human and animal. That is, his ideology seeks separation while his methodology practices translation. Accordingly, until the final catastrophe of the narrative, when a beast person runs amuck and slays its creator, Moreau's lab remains out of sight. A place of nearly total social exile, the experimental bench where these quasi-objects are manufactured remains under wraps, behind blind walls. Before that catastrophe, however, the story brings us to another, somewhat less occulted scene of attempted humanization: the Beast People's commons, where they take it upon themselves to "construct" their humanity through the ritual chanting of social prohibitions. This is one of the narrative's most trenchant strokes: even though religious observances still rival the modern sciences in the vocation of soul making (or "essence production"), the sciences have helped to relativize religions' anthropological stock in trade—the social construction of the human.

An uninvited witness to Moreau's experimental industries, Prendick is no Latourian observer—at least, not at first. What suspense the story generates involves the delay in his realization that Moreau is not animalizing humans but humanizing animals. As in *The Time Machine*, Wells presents a would-be anthropological interpreter whose first attempts to read the status of enigmatic beings within an enigmatic landscape miss the mark. Prendick's "tangle of mystification" is twisted tighter when he encounters "three creatures" performing a "mysterious rite": they "were human in shape, and yet human beings with the strangest air about them of some familiar animal. Each of these creatures, despite its human form, its rag of clothing, and the rough humanity of its bodily form, had woven into it, into its movements, into the expression of its countenance, into its whole presence, some now irresistible suggestion of a hog, a swinish taint, the unmistakable mark of the beast."[34]

Prendick's theological cliché reminds us that traditional ontology had an imprecise label for such confusions of fixed categories: "monstrosity." Under an evolutionary regime, however, this category is recognized as a religious rather than scientific concept. The nonmodern observer sees that monsters do not oppose but rather allegorize the human; they are self-referential projections of the Other within. And Prendick's realization of this, although he resists it and can't process it when it comes, is the crux of the tale. But, once Prendick has comprehended their artifactual origins, he can see Moreau's beast menagerie more clearly as a proliferation of hybrid quasi-objects:

> The two most formidable animal-men were my Leopard Man and a creature made of hyæna and swine. Larger than these were the three bull creatures who pulled in the boat. Then came the Silvery Hairy Man, who was also the Sayer of the Law, M'ling, and a satyr-like creature of ape and goat. There were three Swine Men and a Swine Woman, a Horse-Rhinoceros creature, and several other females whose sources I did not ascertain. There were several Wolf creatures, a Bear-Bull, and a Saint Bernard Dog Man.[35]

The demand placed upon the Beast People to deny their animal origins parodies the moral conflicts of a "humanity" constructed on modern essentialist premises of a human sociality outside of nature, premises that remain tied to the very theological essentialisms the modern Constitution purportedly displaced: "A series of propositions called the Law (I had already heard them recited) battled in their minds with the deep-seated, ever-rebellious cravings of their animal natures. This Law they were ever repeating, I found, and ever breaking."[36] The Beast People have to supplement through ritual communication what Moreau had hoped to accomplish only through surgical reconstruction: the production of the human. In a travesty of religious ritual, the Sayer of the Law chants prohibitions upon bestiality:

> "Not to go on all-Fours; *that* is the Law. Are we not Men?"
> "Not to suck up Drink; *that* is the Law. Are we not Men?"
> "Not to eat Flesh or Fish; *that* is the Law. Are we not Men?"
> "Not to claw Bark of Trees; *that* is the Law. Are we not Men?"
> "Not to chase other Men; *that* is the Law. Are we not Men?"[37]

It seems that Moreau is no Moses—he did not dictate these commandments to the Sayer of the Law or to his People. He tells Prendick: "'They go. I turn them out when I begin to feel the beast in them, and presently they wander there. . . . There is a kind of travesty of humanity over there. . . . I take no interest in them. I fancy they follow in the lines the Kanaka missionary marked out, and have a kind of mockery of a rational life—poor beasts!'"[38] But however the Beast People acquired their totemic tutelage, it is as if the full text of their sayings has self-organized out of communications within the collective of the cast-off "subjects" of Moreau's experiments as they attempt to assume and maintain forms of linguistic subjectivity appropriate to their transformed status as humanized beings.

In their Law the Beast People possess at least an intuition of their scriptedness, that is, of their inscription as quasi-objects within a technoscientific network and thus of the need to maintain their translations into being through the circulation of social communication. This is an aspect of their brief existence that completely escapes Moreau, who, in his difficulties getting the surgical humanization to stick, falls back disastrously on pre-Darwinian notions of fixed essences: "'I have been doing better,'" he tells Prendick; "'but somehow the things drift back again, the stubborn beast-flesh grows, day by day, back again'"[39] If we triangulate Moreau and Prendick from Latour's middle ground, Wells's tale already says that it is their attitudes (like the modern Constitution itself) that are truly anachronistic. What *The Island of Dr. Moreau* says is that the human is essentially nonessential. The human joins the rest of evolutionary life as a quasi-object resulting from a "weaving of morphisms."

For given the capacities of the Beast People for linguistic commerce of human type, it is not Moreau's surgical constructions that have failed. The problem is not ontological but epistemological: What dooms his project is the failure of his own conviction in the status of his results. Moreau has not completed the job of *constructing* his facts: "'These creatures of mine seemed strange and uncanny to you as soon as you began to observe them, but to me, just after I make them, they seem to be indisputable human beings. It's afterwards as I observe them that the persuasion fades.'"[40] Moreau's Promethean blinders cause him, like Victor Frankenstein, to abandon his less-than-perfect creations. And so, to see matters from the other side, Prendick must stumble upon their jungle clearing, where, like homeless street kids, Moreau's creations gather to chant their humanity into being.

The Beast People are woven and further weave themselves from natural and social morphisms. Latour's parallel view of the morphism of the human is a neocybernetic turn putting operational flesh on the bones of the postmodern observation that the human is a rhetorical construction. Indeed, the human lies not in the possession of an essence but in the eliciting and instrumentalizing of a conviction, in a persuasion that it is present—but also, in Latour's terms, in the continuous translation of itself into being by social communications.

Aramis *and the Anthropological Matrix*

Latour's *Aramis, or the Love of Technology* dissects another failed technoscientific project in artifactual animation: the attempted construction and eventual termination of an innovative Parisian "smart" subway network. In that text Latour notes: "There are two models for studying [technological] innovations: the linear model and the whirlwind model. Or, if you prefer, the diffusion model and the translation model."[41] Latour then relates the distinction between diffusion and translation models to one between theological narrative genres implied by these different operational forms. "In the first model, the initial idea emerges fully armed from the head of Zeus," and this yields "a Protestant narrative," or, one of special dispensation or individual election unbeholden to the mediation of networks or collective institutions.[42] Clearly Dr. Moreau is expecting his creations, once they burst forth from his scalpel and operating theater, to save their own dubious souls. "In the second model, the initial idea barely counts. It's a gadget, a whatchamacallit, a weakling at best. . . . In the translation model, there is *no transportation without transformation*," and this yields a "Catholic narrative . . . of incarnation," in which, as it were, the animating spirit can be received only through its repeated translations into wine, wafer, and collective ritual.[43]

I read Latour's sectarian analogues for the narrative exposition of technological networks as translating or desublimating classical spirituality and its residual basis in daemonic metamorphosis, or the ambivalence of the sacred (the potential for spiritual transubstantiations among the bestial, the human, and the divine) into nonmodern daemonic morphism—or, the "variable-ontology world" of the animal, the mechanical, and the human.[44] In this post-Darwinian world, as we noted in the introduction, "the human form is as unknown to us as the nonhuman. . . . It is better to speak of *(x)-morphism* instead of becoming indignant when humans are treated as nonhumans or vice versa."[45]

The metamorphic transformations of bodies—both fictive and artifactual mixings of the human and the nonhuman—recur from archaic to contemporary times, taking daemonic shapes ranging from the magical to the technological. As virtual nonmoderns, despite our modern upbringings, along with Moreau's Beast People we remain within the "ancient anthropological matrix," where "we have never stopped building our collectives with raw materials made of poor humans and humble nonhumans."[46] Textual metamorphs and technoscientific quasi-objects are both mediating transformers performing sociomythic sorting operations, negotiating the relations not of heaven and earth, but of nature and society. Latour's hybrids and quasi-objects, then, participate in a continuous production of ancient and current cultural mediators whose common attribute is a propensity to the metamorphic transformation of given and normative forms. Viewed through the lens of Latour's network concepts, the recursive imageries of literary metamorphoses resonate with the operational evolutions—the mutations and occasional catastrophes—of natural and social systems.

THREE

System and Form

Double Positivity

In autopoietic systems theory the system–environment distinction exhibits a double positivity. There is no shearing off into exclusivities of subject or object, system idealism or environmental materialism. Environments secure the being of systems that secure the knowing of environments in an indivisible loop. Thermodynamically speaking, physical or technological systems may be either closed or open, but the shift to autopoietic (biotic and metabiotic) systems overcomes this antithesis: self-referential observing systems are two-sided: at once operationally closed and environmentally open. Paradoxically, autopoietic closure—the condition of the possibility of self-referential operations—is a positive state that includes what it excludes. Operational closure with environmental openness marks the reentry into the system of the double positivity of the system–environment complex per se. Or again, once the biotic level is achieved or surpassed, autopoietic paradox marks the necessary multiplicity of the total situation.

Luhmann's systems theory poses the challenge of maintaining a rigorous insistence on operational closure of individual systems while also accounting for their operational integration with other systems. When parsing the mechanisms that enable separate systems to interpenetrate without losing their autonomies, multiple positivities supersede dialectical negations and ontological antitheses. Neocybernetic observation privileges differences— the double and the multiple—over unities. Michel Serres has remarked how "every historical era" is "multitemporal, simultaneously drawing from the obsolete, the contemporary, and the futuristic. An object, a circumstance, is thus polychronic, multitemporal, and reveals a time that is gathered together, with multiple pleats."[1] Similarly, for Luhmann, "The phenomenon of meaning appears as a surplus of references to other possibilities of experience and action."[2] A bundle of possible outcomes will be involved in every complex clustering of coordinated systems performances, a bundle that must unfold selectively in the complex time of systems operation. For instance, in the social operation of literary communications, literary observers enact narrative durations—construct readings that decline the virtual simultaneity of the structures of a text into consecutive moments of narrative time. Each moment is an event marking multiple differences: from its previous observer, from the preceding moments of the same observer, and from other possibilities that other observers at other moments will have actualized.

In another doubly positive set of differences, systems theory comes to narrative at the level of systems and the level of forms.[3] The systems level bifurcates into psychic and social references, interpenetrated by the medium of meaning. One attends to the reticulations of operational closure and environmental contingency involved in the ongoing self-reference, reproduction, and evolution of literary systems considered as subsystems in the autopoiesis of social systems. A protologic for the neocybernetic focus on autopoietic systems, the level of forms is the cybernetic level per se. W. Ross Ashby wrote in a textbook of the first cybernetics: "Cybernetics stands to the real machine—electronic, mechanical, neural, or economic—much as geometry stands to a real object in our terrestrial space. . . . It takes as its subject-matter the domain of 'all possible machines.'"[4] Our concern in this and the next chapter is the formal "geometry" of narrative textual structures.

Form cuts across neocybernetic and narrative occasions at an angle that moots many of the current debates over the "new formalism" in literary studies. As Marjorie Levinson depicts these critical arguments, one is asked to choose between sociohistorical or formal–aesthetic concerns, or else between an "activist formalism" or a "normative formalism."[5] As I translate Levinson's analysis, the "new formalists" hope either to recuperate aesthetics for historicism (the activists) or to salvage humanist ideals of the detached literary artifact (the normativists). I would suggest that the double positivity of our neocybernetic approach resolves these issues by enabling interlocking observations of sociohistorical systematics *and* textual formalisms. In the normal course of the autopoiesis of literary systems, literary artifacts are complex structures that circulate (or not) on the basis of multiple receptions of their formal decisions, which at this level also include the constructions of their semantic choices, plot structures, and prosodic schemes. Systems theory, then, in addition to the sociological networks within which narrative objects such as individual texts are inscribed, also grasps the shaping of their machine parts: the particular semiotic patterns that make narrative communications run from moment to moment. "The theory of forms is not yet a theory of systems."[6] Narratives per se are structures, not systems. But as structures enabling effective modes of psychic and social autopoiesis, they form and are informed by the systems that produce themselves by forming them.

History of Paradox

Consider the form of an unsupported circle as a structural metaphor conveying a sense of the paradoxical operations of a self-referential system.[7] A sufficiently long line of people form themselves into a ring and then, on cue, sit down, each on the knees of the one behind them. Here is a structure that could not hold together unless the queue looped back into itself. Were it a mere line of people, the result would be a collapse at its rear end, so to speak. The lack of support for the very last person would cause all the rest to fall down like a row of dominoes. An unsupported circle can hold on the condition that the last person provide support for the first, which can be accomplished by the simple stratagem of forming a circle, that is, by feeding its own form of (non)support back into itself. By doing so, a foundation is created, at least momentarily, out of no foundation. This limns how systems operate: they bootstrap virtual foundations out of circular operations. In circular formations, the distinction of first and last elements becomes insignificant. Once things get going, there is no central element: every one is equally essential, equally defined by its relations to other elements, and all are subsumed into operational unity and formal autonomy by the self-referential circularity of those relations.

Neocybernetic form theory is a kind of grammar that elucidates systems-theoretical propositions. Extended to the milieu of literary theory, it suggests an important revision of narrative formalism. Neocybernetic formalisms describe how the structural autonomy of the narrative text breaks out of iconic fixture and, coupled to an observing system, enters into operational duration. This operational mobilization of semiotic structures can happen, as we will discuss below, because in this way of looking at narrative communications matters of recursivity, self-reference, and paradox are "unfolded" rather than frozen or short-circuited. Systems theory allows the intuitions about reflexivity and paradox dispersed throughout modern aesthetic and literary theory to be collected together and given a rigorous operational explanation and recuperation.[8]

Second-order systems theory is particularly concerned with the circular forms of paradox and self-reference. As we have noted, observing systems display toward their environments a contradictory simultaneity of closure and openness, and neocybernetics investigates the ways that difference and operational closure among biotic and metabiotic systems shape their openness to environments and coupling to other systems. Double positivity is a paradox that infects any and all cognitions, bifurcating them into simultaneous system and environment references. To operate at all, observing systems must elicit and resolve these cognitive issues, bootstrapping themselves out of momentary resolutions of paradox that, always leading to others, keep the system stumbling forward.

Luhmann delineates paradox in the cognitive event—the making/marking of a distinction—that brings any observation about.[9] He notes two traditions of paradox in the West: the logical and the rhetorical. "The logical tradition tries to suppress the paradox" by eliminating the contingencies of observing frames.[10] Classical ontology treats the mind as unconditioned being over and against the nonbeing or phenomenal notionality of its objects. Restated neocybernetically, classical ontology posits the mind as a singular observing system ideally independent of any environment. The crux of its paradox is to be dismissive of its environmental contingencies, blind to its operational (non)foundations. The outsides of its forms are hidden away, and with them the paradoxes or logical binds they would otherwise introduce. But "the rhetorical tradition that invented the term introduced paradoxical statements to enlarge the frames of received opinions— therefore, 'para-doxa'—to prepare the ground for innovation and/or for the acceptance of suggested decisions. . . . The immediate intention seems to be to deframe and reframe the frame of normal thinking, the frame of common sense."[11]

Double positivity demands that paradox not be banished but retained as a productive supplement to other discursive procedures. Thus, for instance, the paradoxes of distinction construct a metalogic alongside Aristotelian orthodoxy by transcending the law of noncontradiction.[12] We generally exchange verbal signs with some confidence that what we and others mean (to distinguish and so indicate) by them is "close enough" to be commonly comprehended. Multiple observations of the same signifier A will converge toward what von Foerster calls an "eigenvalue" or provisional recursive consensus. Still, especially if autopoietic systems and their observations are being referred to, A at one moment in time cannot be entirely identical to A at another moment. At bottom every linguistic (or literary) observer constructs a different version of A. Thus $A \neq A$—"the same is different." Literary narratives of metamorphosis induce a comparable paradox of shifting frames: when the body changes form but the person of the character doesn't, "the same is different." The narratives generated by such paradoxes join the tradition of rhetorical paradox in just the role Luhmann cites—to train the observer to exploit disorientation (informatic noise) through the supple substitution and complication of cognitive frames.

Systems theory approaches literary objects and events by analyzing the paradoxes latent or patent in figures of narrative observation. But in systems theory the concept of observation is generalized beyond our common assumptions of conscious, sensory, or perceptual surveillance. Abstractly considered, observation occurs whenever and in whatever way a state of things generates or receives a mark. To observe is to mark a distinction, in some traceable form, an inscription that supplements as well as indicates the state it marks. Observation is a condition of the viable operation of systems: cells, brains, psyches, and societies each have ways of marking and processing distinctions to maintain their particular systemic functions. Autopoiesis is the temporal, ongoing self-organization of systems through self-referential operations guided by environmental observations.

It may be assumed that premodern religions, philosophies, and literatures have registered insights into these regimes of biological, psychological, and social organization. Under other vocabularies and figural regimes, the paradox of observing systems is rewritten as the sacred or the daemonic. Myths of psyche and society consistently show that openness to the other (effective cognition of the environment) depends on the self-referential closure—the identity or integrity—of the system that knows. Yet autopoietic systems' operative closure—their systematicity—is coupled to their openness to environmental contingencies that in turn continuously reorganize the system. Meaning systems in particular—psyches and societies—self-organize through the interplay of auto- and hetero-reference, self-observation and environmental registration. By this definition, to exist and to function, autopoietic systems must produce those forms of distinction that enable these operations of observation. For instance, in stories that turn on metamorphic transformations, the play of metamorphoses can be described as a sequence of narrative operations that shows forth the systematic effects of reversible crossings over the formal boundaries of various conceptual distinctions.

Laws of Form

Luhmann couches his later theory in a vocabulary derived from British mathematician George Spencer-Brown's *Laws of Form*.[13] This "very unusual and contentious work" is a primary thread tying second-order cybernetics together from its early formulations in von Foerster to the development and extension of the concept of autopoiesis in the work of Varela and Luhmann.[14] *Laws of Form* presents a calculus of distinctions that systems theory has correlated with marks of observation.

Distinctions result when observations leave a mark. Systems observe by marking distinctions and crossing over between marked and unmarked states. The marking of a distinction produces the following elements, collectively referred to as "the form":

- the indication, the marked state: "the inside the distinction"
- the indication's exterior, the unmarked state: "the outside of the distinction"

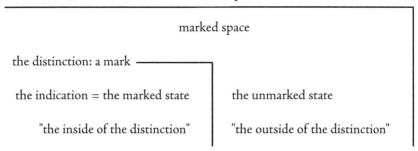

Figure 3.1. The Form

- the distinction itself as the unity of marked and unmarked states
- a second distinction between marked and unmarked spaces

A single act of distinction always already produces another distinction from which an infinite series of distinctions can ramify—and conversely, a complex of systemic distinctions can often be collapsed back into (the multiplicity of) a single observation. A distinction also marks, in addition to the indication, the distinction itself. Any mark marks itself as well as that which it marks. American mathematician Louis H. Kauffman develops Spencer-Brown's calculus of form in noting: "A mark or sign intended as an indicator is self-referential. . . . The mark refers to itself. The whole"—that is, the entire complex of marked and unmarked states and spaces initiated by the mark, the totality of the form—"refers to itself through the mark."[15] Thus the operations of systems depend on observations that are elementally self-referential, and the systems built from such elements are systematically self-referential.

At any given moment in the operation of an observing system, distinctions can function only on the inside of the forms they produce. This finitude of observation in time is the formal correlate and the burden of autopoietic closure: what a system can know of its environment is necessarily translated into and maintained in the medium of its own elements. Cognition is primordially self-referential: "only closed systems can know. . . . [This is] a paradox: it is only non-knowing systems that can know; or, one can see only because one cannot see."[16] An observing system can see only the inside of the distinctions it makes at any moment of its operation. It cannot simultaneously see the unmarked state of those same distinctions—it cannot form what Luhmann denotes the "unity of the distinction," or it can do so only in the form of paradox or logical contradiction. Complex systems generate time because they need time in which to unfold the paradoxes inscribed in observing forms.

Distinctions are paradoxical because they imply a coexistence between their inside and outside terms that cannot be made operational by the observing system using them at the moment of their use. In Spencer-Brown's form notation, this is indicated by the location of the indication within the enclosure defined by the concavity of the distinction mark. In his calculus of distinctions, the reiteration of the distinction (for instance, to repeat someone's name until they hear you calling) refreshes but does not alter the form of the observation. "The value of a call made again is the value of the call."[17] Spencer-Brown's "law of calling" has the "form of condensation":

Figure 3.2. The Form of Condensation

To access the remainder of the information carried by the total form of distinctions, the system must cross over the initial distinction to mark its outer side, which crossing unmarks the previous indication. "For any boundary, to recross is not to cross."[18] Spencer-Brown calls this the "law of crossing," or the "form of cancellation."

Figure 3.3. The Form of Cancellation

Yet this clearing of the marked space is not an emptying: it is the deconstruction of the initial construction, which frees the system to re-mark the mark, to maintain it as an ongoing observation, or perhaps, to introduce a different indication the next time around. Observation can operate only while momentarily blinded to the paradoxicality of its own forms. Social systems communicate through complex oscillations between the calling and the crossing, the condensation and the cancellation of shared indications.

Paradoxes in the forms of observation can be resolved—that is, rendered productive—by another observer's observing the initial play of distinctions. This can also be diagramed as a form of cancellation, but in this case expanded to indicate a second-order observer taking as an indication the entirety of another observer's distinction.

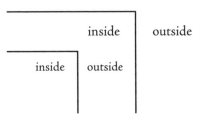

Figure 3.4. Second-Order Form

Any one observer in isolation is no more than a psychic system; two observers are the minimal beginnings of a social system. A woodcut by Albrecht Dürer helps clarify this distinction between first-order and second-order observation by placing its viewer in a second-order relation to the depicted first-order observer. Figure 3.4 presents a clearer diagram of Dürer's image if you rotate the second, outer distinction mark diagonally to the first, aligning it with your view of the draftsman's view.

In first-order observation (here, the draftsman as first-order observer of the model) all that can be seen of the form of observation is the indication, the inside of the distinction (here, the model that will occupy the entire frame of the drawing being made). The draftsman himself will not appear

Figure 3.5. Albrecht Dürer, from *Unterweysung der Messung*, Nuremberg, 1538

or be explicitly manifest anywhere in that drawing. In other words, in first-order observation the outside of the distinction is thrown into a cognitive blind spot. Moreover, this image captures the frame (see chapter 4) precisely as a technology of observation. In fact, perceptions are complexly framed by the different sensory apparatuses of the psychic system doing the perceiving, although these contingencies, because habitual, are generally unconscious. In Dürer's image we can clearly see how the draftsman's viewing net literally frames the object: it participates in its reconfiguration as a foreshortened image rendered through "Renaissance perspective," then disappears. In addition, once the image is processed through that portal, nothing of the body depicted remains either. Both the apparatus of distinction and its prior indication have been erased. This is itself an allegory of the larger patriarchal scenario enacted here, in which the male gaze sighting along a phallic index takes in a female object only to reduce it, according to the prescriptions of single-point perspective, to a shadow on a page.[19]

But the form of first-order observation feeds back into itself when one observes the operations of another observer's observation. In the second-order observation of first-order observations, the whole form of the distinction that produces the first indication—the "unity of the distinction" between that which is indicated and that which is not—can now be observed, precisely as Dürer has done in this perspectival study of the shifting and multiplying of perspectives. Here what is seen is not just the object but also the formal (and ideological) mechanisms of its observation: Dürer's second-order perspective renders observable the way that the initial object is being observed. Of course, the second-order observer is always also in a first-order position, blind to or occulting the environment of its own system of observations. But here again, although the formal conditions of observation are paradoxical, they work—as long as one has time to keep shifting perspectives as well as submit them to another's construction.

In his *Whole Earth Catalog* review, von Foerster touched on a key point in the *Laws of Form*, unfolding implications for the recursivity and temporality of systems that Varela and Luhmann will later capitalize on for biological and social systems theory. This is the concept of *reentry*: "In Chapter 11, Spencer Brown tackles the problem of infinite expressions by allowing an expression to re-enter its own space. This calls for trouble, and one anticipates now the emergence of antinomies." That is, the function of reentry is

analogous to the production of logical paradox or antinomy by the introduction of self-referential propositions into a syllogism. But although one
had expected reentry to produce paralysis through infinite regress, von
Foerster is delighted to note that within Spencer-Brown's calculus of distinctions: "Not so! In his notation the classical clash between a simultaneous Nay and Yea never occurs, the system becomes 'bi-stable,' flipping from
one to the other of the two values as a consequence of previous values, and
thus generates time! Amongst the many gems in this book, this may turn
out to be the shiniest."[20]

For any observing system, reentry is the process by which its environment, the outside of the system, is registered on the inside of the system.
In other words, reentry is the formal hinge that joins autopoiesis as self-
production to autopoiesis as cognition. Reentry is the crux of the constructivist epistemology demanded by rigorous theories of organizational closure, from Varela's insistence that an autopoietic system has no inputs and
no outputs, to Luhmann's statement that "only closed systems can know."[21]
In a self-referential system, all reference beyond the system will necessarily
be self-constructed by and within the system. Through the form of reentry,
*the system constructs within itself as a virtual reality the real distinction between
itself and its environment.*

Systems that lock onto mere repetitions of the same eventually spin out
of existence. Literature in particular is an epistemological device for inter-
minably deferring the location of an ultimate perspective from which the
being of things could be thought to be known once and for all. The cognitive effects of narrative and rhetorical structures rehearse, both dramatically
and thematically, the framing and deframing of paradox that neocybernetics
observes in cognition per se. Narrative transformations can be read as an
allegory of the incessant temporal dissolution and reconstruction of forms
within autopoietic systems.[22] Literary metamorphoses replicate the incessant protean restructurations of viable systems, of constant fluxes within the
medium of elements out of which forms can be assembled and observations
inscribed—in the case of psychic and social systems, the medium of meanings. The dissolutions necessary to the maintenance of cognitive formations
are usually occluded or rendered unconscious by the psychic system's very
operation, as consciousness remains on the inside of its immediate form.
The paradox of observation might be brought down to this statement: *non-
knowing frames the ability to know.*

Form in A Midsummer Night's Dream

Von Foerster's dictum that paradox "is good for you, if you take the dynamics of the paradox seriously," is borne out by the metamorphic comedy of *A Midsummer Night's Dream*. Paradoxes of observation are given symbolic form when cognitive reversals or cultural antinomies produce contradictory or discordant images or narratives. What is new is the neocybernetic way of observing them. For example, we can say that *A Midsummer Night's Dream* unfolds and refolds the paradox of observation by reframing human eros within and without a visible–invisible fairy embrace. With passing reference to this drama, David Roberts has developed a number of the implications of literary and dramatic form for neocybernetic analysis: "The invisible boundary, that is, the invisible distinction that separates audience and play but also play and world, is made visible through the re-entry of form. . . . The play within the play can manifest within itself . . . the crossing of the boundary between the inside and the outside of form, between illusion and reality, play and world."[23] In literary terms, then, the form of second-order observation can be folded back into the structure of a story through the formal interplay of outer and inner—diegetic and metadiegetic narrative levels or dramatic scenes. Second-order observation emerges, for instance, when a story is told within a story, and we observe both the story and its teller simultaneously. Famously, in act 5, the "very tragical mirth" of the Pyramus-and-Thisby skit reenters the oxymoronic form of this metamorphic comedy back into the play. And too, at the very end, Puck also inverts the frame of the play, turns to its audience, observes his observers, and advises them how to construe the play's constructions.

Nonknowing frames the ability to know. A Midsummer Night's Dream is pervasively constructed around this very epistemological paradox, as articulated by Helena in the first scene: "Love looks not with the eyes, but with the mind, / And therefore is winged Cupid painted blind."[24] In the end, the lovejuice brewed by Oberon and administered by Puck does not cloud but rather makes transparent the blind spots that enable us to see, and to see each other. *A Midsummer Night's Dream* is built upon the observation of observation, and so offers some vivid dramatic formulations of these basal paradoxes of distinction. The play revels in the transformations of identity produced by shifting observational frames, and bodies forth the ironies and inversions of form in the shape of dream beings that exhibit the paradoxical unities of the distinctions that structure the waking world. Narrative embeddings and chiastic structurations interpenetrate this drama's panorama of metamorphic changes.

Neocybernetic observation would underscore the interpenetration and coproduction of psychic and social systems, which mutually compensate for the other's forms of closure. Closed but environmentally contextualized, recursive systems are the fairy "ringlets" caught in "quaint mazes," the mundane miracles that make things happen, or else break down from noncommunication, like the union of Titania and Oberon. Their play communicates the interplay of social and psychic systems by staging erotic miscommunications through the interaction of two closed yet interpenetrating societies, human and fairy. The play is an allegory of art, which in its operation as a social subsystem "integrates perception and communication without merging or confusing their respective operations. Integration means nothing more than that disparate systems operate simultaneously (are synchronized) and constrain one another's freedom."[25]

In *A Midsummer Night's Dream*, observation is embodied in the agency and behavior of dramatic characters and denoted as "marking." To observe is to mark—to inscribe a mark of distinction—as when the fairy queen Titania speaks to the fairy lord Oberon about the late mother of the changeling boy at the center of their marital dispute:

> His mother was a vot'ress of my order,
> And in the spicèd Indian air by night
> Full often hath she gossiped by my side
> And sat with me on Neptune's yellow sands,
> Marking th' embarkèd traders on the flood.[26]

In fairy allegory, Titania recalls a shared moment of visual observation, marking out or indicating an object as distinct from the unmarked or undifferentiated flux or "flood" of perceptions. To this first-order observation Oberon is then brought forward as a second-order observer; that is, he enjoys a perspective from which he can observe Titania and mark the transformations of her observation when under the influence of the lovejuice. In the scene at hand, Oberon distinguishes the Indian votaress from an "imperial vot'ress," presumably the virgin Queen Elizabeth, whom Oberon observed eluding the "young Cupid's fiery shaft":

The imperial vot'ress passèd on,
In maiden meditation, fancy-free.
Yet marked I where the bolt of Cupid fell:
It fell upon a little western flower,
Before milk-white, now purple with love's wound,
And maidens call it love-in-idleness.[27]

Systems observe by marking crossings over distinctions between marked and unmarked states, or between first- and second-order perspectives. The play of implied distinctions continues here with the difference between Titania's horizontal or first-order purview from the Indian seashore and Oberon's vertical or second-order god's-eye view, from which unspecified vantage only he can see, "flying between the cold moon and the earth / Cupid, all armed."[28] Titania makes a maternal metaphor—the way that "the sails conceive / And grow big-bellied with the wanton wind"[29]— Oberon marks the flight of a phallic shaft deflected from the Virgin Queen to a surrogate virgin, a flower that is now "deflowered" and transformed from white to purple. This "marking" now distinguishes between symbols of sexual innocence and experience, and the formal crossing of this particular, irreversible distinction between maiden and nymph, virgin and mother, is a kind of paradigm for the metamorphic business soon to follow, precisely with the juice of this transformed flower as Oberon's *pharmakon* of choice.

A Midsummer Night's Dream is rife with forms of reversible observation. The primary form of its mise en scène is the distinction between the human and fairy worlds.

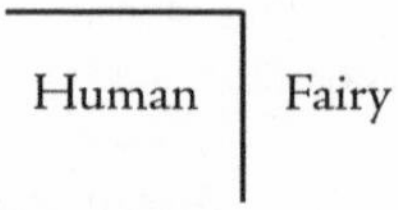

Human | Fairy

Figure 3.6. The Human/Fairy Distinction

Act 1 begins on the inside of this distinction, with the human characters absorbed in their all-too-human problems. Act 2, however, immediately recrosses the human–fairy distinction and displays the fairy world, renders it visible by an audience placed at a second-order level of observation. The fairy world is now the inside of the distinction, while the human world is shown to the audience as the environment that fairy society is embedded in. This chiasmus or amplification through reversal of perspective underwrites all the other transformations to follow.

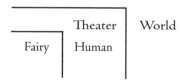

Figure 3.7. The Theater as Second-Order Observation

As we can see, the observability of the fairy world is predicated on the audience's remaining on the inside of the theater: to recross that boundary is to dispel the "magic." But within the spell of the play, once the human characters cross over from the city to the countryside, it is as if they have also crossed the boundary from the real world to the scene of a play. Or again, the human crossover into the fairy world implicitly reenters the form of the play into the play, in a manner to be made explicit by the play within the play that will be performed once the human cast crosses back to the world of the court. In short, the fairy world is an allegory of the theater, just as the theater is an allegory of the world. Puck, the daemonic metamorph at the threshold of the stage, is an allegory of dramatic art. Art per se embeds itself with intuitions, then embeds itself within society.

As a prologue to that festive, self-referential reinscription of the play back into the play, Shakespeare gives his most elevated human characters a famous passage of metacommentary on the dramatic actions of the midsummer's night.[30] This passage has always been taken, in addition, as an observation on the paradoxical nature of drama and literary fantasy altogether. A first-order observer and psychic isolationist, Theseus cannot mark the outside of the human nor the inside of the fairy indication, so he sees only "airy nothing" in the collective hallucinations and "fairy toys" of the midsummer-night lovers. Hippolyta peers behind his blind spot, observing at second order by speaking with the lovers and having them renarrate their "dreams," retelling over "all the story of the night." Hippolyta is an unabashed epistemological constructivist: by adding all the dream accounts together rather than subtracting them from the real, she constructs "something of great constancy." Or as we might say, she intuits an eigenvalue, something systematic about the whole affair.[31] Comic reunions and climactic weddings bring about recursive regularities in the communications among "minds transfigured so together." Neocybernetically speaking, comedy's aim is to maintain the autopoiesis of social systems.

The daemonic and metamorphic elements of *A Midsummer Night's Dream* may be taken as further reifications of the form of observations. The metamorphic Bottom and the daemonic changeling boy unfold the paradoxes of linguistic, sexual, familial, and social distinctions. These characters are then inserted into embedded frames that elicit from the structures of dramatized observations the formal conditions of the paradoxical spectacles they contain. The play of formal reversals that generate the metamorphic elements of this play is marked in particular within the dramatic dialogue by a recurrent rhetorical scheme: the figure of chiasmus. X marks the spot where a chiasmus enacts a self-referential linguistic reversal. By looping a sequence of elements back upon itself, the scheme of chiasmus reenters the form of its initial phrase back into itself the other way around. In this way the rhetorical crossing of a chiasmus also reenters the dramatic crossings of formal boundaries back into the text, embedding the text with miniature emblems of itself. Bottom registers this mise en abyme of self-referential levels in the structure of the drama in his reflexive self-commentary that his dream experience "hath no bottom."[32]

Chiasmus inscribes this text with formal re-markers of "crossings," as when Hermia and Lysander reflect on the way that their star-crossed plight typifies lovers' destinies: "If then true lovers have been ever crossed . . . it is a customary cross."[33] Later dialogue makes this implicit chiasmus explicit by describing chiastically the crossing of realms intrinsic to the daemonic function per se, that is, the crossings inscribed by agents of metamorphosis as well as by the metamorphs they create by crossing. The thrust of this chiasmus, however, is first elicited in this same give-and-take between Hermia and Lysander, when he reflects how the happiness of lovers—the duration of their idyll before destiny comes to cross it out—is typically "brief as the lightning in the collied night / That in a spleen unfolds both heaven and earth, / And ere a man hath power to say 'Behold!' / The jaws of darkness do devour it up."[34]

Here, it is the lightning flash itself that illuminates and thus makes observable both the alacrity of dissolution in systemic operations and the gap between heaven and earth, the human and the divine, in which medial space daemonic agents such as Cupid wing their way. That the scheme of chiasmus underwrites the form of metamorphosis in this text is further underscored at the end of the first scene, when Helena in soliloquy remarks about Demetrius that "herein mean I to enrich my pain, / To have *his sight* thither and back again" (my italics).[35] This locution connects the transformation of observation to the figure of chiasmus per se as a looping motion that moves "thither" only to recross itself "back again," and to the action of the drama about to commence as the human players move "thither" into the dream enchantments of the midsummer's night and then "back again" into the waking world. Once that boundary is recrossed in the aftermath of the fairy-nocturnal episodes, Shakespeare unfolds the total chiasmus or antimetabole at play in the famous speech of Theseus. Theseus cannot see what he cannot see—that his chiasmus "from heaven to earth, from earth to heaven" (which completes the rhetorical circuit opened by Lysander's lightning bolt) describes the formal crossings of observation by "the poet's eye" that transform the "airy nothing" of forms into substance. But what his words say produces "something of great constancy":

> The poet's eye, in a fine frenzy rolling,
> Doth glance from heaven to earth, from earth to heaven;
> And as imagination bodies forth
> The forms of things unknown, the poet's pen
> Turns them to shapes and gives to airy nothing
> A local habitation and a name.[36]

In brief, the metamorphic turnabout of the drama at large moves from *the transformation of observation*—as when fickle male lovers turn their erotic gaze from one female to the next—to *the observation of transformation*. Within the play, this prerogative or privilege to observe transformation is reserved in the fairy night to the lord and liege of the fairies, and on the human wedding night to the lords and ladies of the court. Oberon sees Cupid wound the "little Western flower" and observes its transformation, then uses the juice of love-in-idleness to transform the observations of lovers and asses. Just as the invisible fairies observe the laughable humans, it is given to the most laughable human—the transformed Bottom—to observe the viewless delights of the Fairy Queen. The chief metamorph of the play, Bottom, becomes a hybrid monster that visibly displays the unity of the human–animal distinction while entering into the unity of the human–fairy distinction. Then, in act 5, he undergoes dramatic transformation into the doleful Pyramus, and reenters the form of the play once more into itself, not just by setting forth the play within the play, but also within the skit by continually breaking its frame. He crosses and recrosses the distinction between the embedded drama's fictive indications (self-reference) and its audience (other-reference), as Puck then does from within the fairy frame that now encloses and is enclosed by the human stage. The form of metamorphosis in *A Midsummer Night's Dream* observes the art of drama as a loosening of the elements of psychic and social systems—their given and projected environments—enough to turn them into media for poetic and autopoietic transformations.

Narrative Forms

I. EPISODES AND LEVELS

Restated as a second-order systems phenomenon, narrative is a form of communication taken from and realized through processes of observation that arise within social systems of sufficient cognitive complexity. Who is to say (since we are disbarred from entry into their society) that songbird couples calling in simultaneous harmony, or whales threading the deep with rhythmic cries, are not communicating socially processible stimulation in tandem with their mates, producing proto-narrative scripts for further observation and operation in those nonhuman societies? Narrative texts are complex structures, macromolecules of communication, which are processed, selected for or against—discussed, reproduced, translated, and so forth, or not—by the autopoiesis of social groups. Let us return to formal matters by redescribing some narratological forms, this time in the vocabulary of neocybernetic form theory. The goal is to shape neocybernetic formalisms capable of coupling together the *inscription* (textual structure) and the *operational processing* (perceptual and communicative systematics) of narrative operations and observations, to map more rigorously their temporal complexity.

The inexorable trend toward observing narrative as a systems phenomenon is sketched by the contrast between two famous graphic models of episodic structure in the fabula: Freytag's 1863 "triangle" and Bremond's 1970 "four-phase cycle":

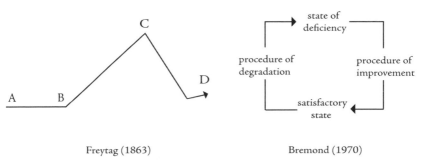

Freytag (1863)   Bremond (1970)

Figure 3.8. The Freytag Triangle and the Bremond Cycle[37]

Freytag's model is really not a triangle—the line it traces does not return upon itself, but rather gives the cross section of a propagating wave. As applied to a novel, for instance, Freytag's open triangle may be viewed as a fractal wave of successive episodes that are self-similar in affective detail (exposition, "rising action," denouement) to the narrative arc as a whole. With Bremond's model, however, late classical linearity morphs into cybernetic recursivity—a "cycle" generated by the constant re-inputting of output. Specifically, Bremond's model is a diagram of narrative homeostasis, an icon of first-order cybernetics. Both models indicate that the propagation of narrative structures is sustained by some cascading or reiterating concatenation of narrative distinctions. These diagrams are focused in particular on the sub-borders or brackets that demarcate narrative *episodes*: virtual boundaries constructed by functional distinctions in successively different narrated events, usually bound together in some fashion by the unfolding rhythms of (ana)chronological events.

Another standard starting point for narrative theory is the notion of narrative *level*. In Gérard Genette's technical terminology, this is the *diegesis* of a narrative, a jargon we have been lessening with the term *storyworld*. In systems jargon (which we cannot avoid), the storyworld is the narrated environment constructed by the discursive inscriptions of narrators and focalizors whose productions imply the operation of observing systems. The storyworld as narrated environment can be known only as the construction of a narrating system. The narrating system and its environment are either distinct from or a part of the storyworld produced by the narration. In other words, what Genette calls an extradiegetic narrator typically resides in an unmarked environment, within which the storyworld is embedded as the marked space. Alternatively, an intradiegetic narrator is already embedded as an observing system within the storyworld produced by its own narration. The system–environment distinction, as described through the double positivity of the *Laws of Form* calculus, gives a formal–operational inflection to Genette's insightful description of diegetic multiplicity: "*Any event a narrative recounts is at a diegetic level immediately higher than the level at which the narrating act producing this narrative is placed*" (italic in original).[38]

By processes internal to the virtual system that produces them, narrators reconstruct narrative levels (Genette's *diegeses*)—story–fabula environments they indicate as external to the (narrating) system. The emergence of narration at all indicates some location in narrative space, a storyworld the conditions of which each narrator will specify (or not). Narrations proceed at a particular level, but are always observed from a virtual level that subsumes it. Levels of observation can always differ because any observation can be embedded in another's observation. In observation in general and narrative in particular, one is always already given more than one observing system (at the least, the author–narrator distinction) and more than one environment (at the least, the inside *and* the outside of a diegesis).

Observational multiplicity also ramifies within the same observer—and so, within the same narrator—as the distinction between narrating system and its narrative environments reenters the system in question. Oscillations among separate observers are joined by oscillations among different situations of observation within the same observer. Casual perceptions, say, of the scenery going by on a road trip, differ from complex cognitions: observations concentrated by reflection on their own operations. In the visual instance, here you watch how others watch and consider what they see according to the way they look at it. Or you may watch yourself watching to determine that you are watching (yourself or others) in an effective or productive way. The distinctions between non-reflective and reflective attention, and between self- and other-observation, are variants of the neocybernetic distinction between first-order and second-order observation. And the form of that distinction aligns itself with other basic forms of narrative theory, in particular, narrative *situations*.

2. NARRATIVE SITUATIONS

Systems–theoretical form theory is the methodical, prelogical unfolding of related intuitions developed around the phenomena of systemic observation, framing, and embedding. Thus its close fit with the structures of narrative situations, and its potential to clarify their operational as well as structural character. Narrative situations are "rough descriptions of basic possibilities of rendering the mediacy of narration."[39] Narrative situations are also *frames*: forms of observation defined by the placing and maintenance of cognitive borders. Their oscillation, or not, in the course of a narrative sequence is a coding device of the first order. In Stanzel's useful vocabulary, narrative situations reflect a basic distinction between *character-bound* ("first-person" or "intradiegetic") and *authorial* ("third-person" or "extradiegetic") narration. But this traditional distinction bifurcated again once authorial narration was seen to be capable of oscillating in its own right between two distinct modes: the strict authorial and figural situations, that is, between indirect ("overt extradiegetic") and *free* indirect ("covert extradiegetic with internal focalization") discourse.

One can give these narratological concepts a rigorous neocybernetic construction by redescribing them through the *Laws of Form*. Spencer-Brown's calculus of distinctions allows one to diagram the situational unfolding of narratives in process, and *as* processed or reconstructed at the receptive end of the narrative communication. In Stanzel's setup, after the foundational cut between author and narrator, the next cut is between two kinds of narrator: character-bound and authorial. This distinction is then reentered on the side of the authorial narrator in the two kinds of authorial narration: strict authorial and figural. In every case, the distinctions depend upon the narrative text's calling and crossing of operational boundaries. To begin with, by definition, a character-bound narrator (CbN) resides within the storyworld it is narrating into being. A naive sketch of this narrative situation from an observational vantage could look like this:

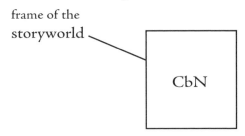

Figure 3.9. The Intradiegetic Status of the Character-bound Narrator

Let us simplify this diagram with form notation:

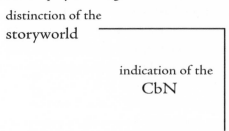

distinction of the
storyworld

indication of the
CbN

Figure 3.10. The Character-bound Narrative Situation in Form Notation

Character-bound narrators are virtual observing systems immersed within but operationally distinct from their narrative environments: they cannot perceive all and thus communicate for certain only that which is turned toward their perceptions of that world, that which they are in a position to turn their perceptions toward. In contrast, while character-bound narrators are contained within the frame constructed by their own perceptual enclosures (by the operational closure of observing systems), the authorial narrator (AN) stands outside the storyworld of its characters:

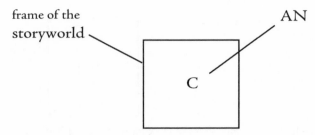

frame of the
storyworld

AN

C

Figure 3.11. The Extradiegetic Status of the Authorial Narrator

Restated in form notation:

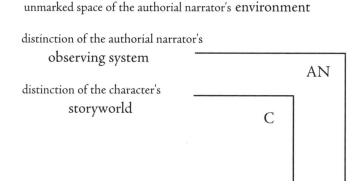

unmarked space of the authorial narrator's environment

distinction of the authorial narrator's
observing system

distinction of the character's
storyworld

AN

C

Figure 3.12. The Authorial Narrative Situation in Form Notation

The authorial narrator operates a *cancellation* of the form that cuts the characters' world out of the unmarked space. It can recross the frames of the characters' storyworlds, as well as move across the boundaries of characters' psychic systems within those storyworlds, observing them from both sides, not at once of course but in successive moments as the narration unfolds. Or again, like Oberon in *A Midsummer Night's Dream*, the authorial narrator occupies the unmarked space of the characters' storyworld, and can see both sides of the distinctions they deploy as first-order indications. With the advent of an authorial narrator the operational closure of the character's frame is superseded as far as the narrator is concerned, and the narration as a whole now finds its limit in whatever binds the narrator's horizons. Of course, narrators are under no obligation to draw a map of their own environment, but interpretive attention can be usefully applied to discerning the virtual borders implied by a narrator's range of observation. Obviously, again, the total environment of narration is finally unbounded and exceeds anything any narrator could cognize. Nevertheless, to the extent that an authorial narrator does cognize some extradiegetic environment, the construction of that world will be delimited at some point by the finitude of its possible observations.

We have been discussing authorial narration in the strict sense defined by its immediate differentiation from the character-bound situation. In neo-cybernetic terms, the *figural* mode of the authorial narrative situation, free indirect discourse, arises when the second-order form of authorial narration is reentered into the authorial situation. One could call this "third-order" narration, but that would obscure the fact that the boundaries of the figural narrative situation remain set by the authorial narrator's horizon. So we will continue to call this a variation—a differentiated recursion—of second-order authorial narration. The key is that, in the figural narrative situation, authorial narrators seem to abscond from their own narration (render themselves "covert") by straddling or merging with the two-sided frame between their narrative level and the storyworld over which they narrate.

The narrative situation of Thomas Pynchon's *The Crying of Lot 49* is largely authorial, with an intense primary focalization through the perceptions of main character Oedipa Maas. The following paragraph from a well-known passage begins in the strict authorial, "omniscient" extradiegetic situation. The narrator performs an external observation of Oedipa's internal discourse with herself from the observation deck of an all-night bus going in circles around Los Angeles:

> Last night, she might have wondered what undergrounds apart from the couple she knew of communicated by WASTE system. By sunrise she could legitimately ask what undergrounds didn't . . .[40]

The continuation of the passage, however, shifts to the figural situation by crossing back over the border of the character to place us inside the frame of Oedipa's thoughts so they seem to narrate themselves:

> . . . what undergrounds didn't. If miracles were, as Jesús Arrabal had postulated years ago on the beach at Mazatlán, intrusions into this world from another, a kiss of cosmic pool balls, then so must be each of the night's post horns. For here were God knew how many citizens, deliberately choosing not to communicate by U. S. Mail. It was not an act of treason, nor possibly even of defiance. But it was a calculated withdrawal from the life of the Republic, from its machinery. Whatever else was being denied them out of hate, indifference to the power of their vote, loopholes, simple ignorance, this withdrawal was their own, unpublicized, private. Since they could not have withdrawn into a vacuum (could they?), there had to exist the separate, silent, unsuspected world.[41]

No wonder Oedipa is paranoid! When the authorial narrator shifts to figural mode, the system–environment of the narrator intrudes into the storyworld of the character, and the machinery of discursive manipulation is both laid bare and spirited away. Hence the postmodern situation of metafictive characters reflecting upon their narratedness in the midst of the narration. Pynchon's narrator does not press Oedipa that far; nevertheless, in this passage the narrator reinforces its implied narratee's cognition of the figural attitude with the quick parenthesis "(couldn't they?)," to indicate that the passage about to conclude is still to be read as Oedipa talking to herself.[42]

A diagram of authorial narration in the figural situation could look like this:

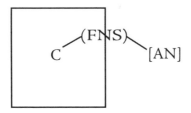

Figure 3.13. The Figural Narrative Situation

Restated in form notation:

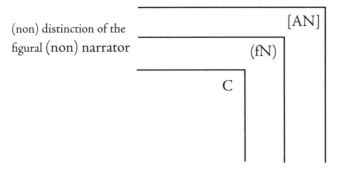

Figure 3.14. The Figural Narrative Situation in Form Notation

An important point about the figural narrative situation is that there are, properly stated, no figural *narrators*: there are only authorial narrators that lapse from strict externality, that go from overt to covert to defer the attribution (but not the operation) of narration to a character internal to the storyworld. That is why figure 3.14 distinguishes a (non)narrator within the distinction of the authorial narrator. But the figural (non)narrator occupies the position of narrative mediation per se, and so indicates the virtual mediality—Jakobson's phatic function, the absconded materiality of communication—that always already supplements relations between senders and receivers, subjects and objects, narrators and storyworlds.[43] As we have said, in the figural narrative situation, the storyworld has not been enlarged because the external cognitive boundaries of character and authorial narrator have not changed. Instead, those boundaries have imploded *on the inside*—more precisely, internal character and external narrator have bootstrapped themselves together to reduce entropy with an increase in complexity.[44] The storyworld in the figural narrative situation *has* become more complex, more infolded. The space of the storyworld has become thicker with a kind of distributed cognition, as a narrating system external to the storyworld is entered into the storyworld through a character's internal observations.

The figural narrative situation marks the discovery and exploitation of *two-sided narrative form*. These narrative techniques for virtual voicing first emerged in nineteenth-century fictions (Austen, Flaubert), then achieved critical mass in modernist novels (Joyce, Lawrence, Woolf). Free indirect discourse was thus more than ready for action when cybernetics and informatics further imploded the infolded space of modernist representation into the ethers of semiosis and cyberspace. The calculus of distinctions would compute the notation of the figural narrative situation, by the form of cancellation, so as to leave one distinction mark standing:

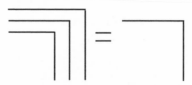

Figure 3.15. Form Calculation of the Figural Narrative Situation

At any given moment, the indication of the distinction left standing could be any of the three participants to the narrative calculation: character, authorial narrator, and/or the form or boundary of their mediation. Figural narration gives form a self—by which we can observe the constructedness of narrative selves at any level. The form of figural mediation is a supplement both to the character and the strict-authorial narrator. The two-sided

distinction that marks the figural narrative situation instantiates the authorial narrator without and within the mind of the character. The character
speaks, but does not speak; the character has ideas of which he or she or it
has no idea. If strict authorial narration is "authorial omniscience," the figural narrative situation is "authorial telepathy."[45]

The figural narrative situation is only one formal step away from literary
fantasies of literally telepathic characters, or, more recently, telematic characters narrated as throwing their thoughts and virtual bodies around in
cyberspaces. The "powers" of such characters result from *character-binding*
the figural situation of authorial double-framing. By granting or technically
equipping a character or character-narrator with the two-sided, second-
order powers of an authorial narrator, that character is transformed into a
mind reader. Once again, now from the side of the character, the cognitive
frame of perceptual closure is crossed out. Narrative forms that post the
human beyond its systemic boundaries never go out of vogue with linguistic
subjects. The receivers of stories never seem to tire of the pleasure that
narratives provide by fulfilling desires for the virtual reality of overcoming
the operational closure that binds all real observing systems.

Two-Sided Form

Luhmann argues that the concept of form is underconceived if taken only
as the antithesis of some counterconcept such as substance or matter, something that is essentially not form, something formless unless supplied with
form. That is, the classical form of form is *one-sided*: especially as framed by
the being–nonbeing distinction, form is distinguished from a more or less
formless and thus ontologically marginal opposite—form/substance, form/
matter, form/content.[46] But under classical dyadic logic or dialectical antithesis, the form term always creeps back into the nonform term. Deconstruction already shows us this much. Discontented with that logic, and with
the limitations of these typical dichotomies and the imprecision of their
associated classical formalisms, other postmodernisms have sought a radical
recuperation of the antithetical term in the valorization of entropy, liquid-
ity, or the formless.[47] Systems theory offers a further and arguably more
powerful resolution of the form of form: form is reconceived as the *two-
sided boundary* by which some thing or thesis is differentiated from something else. The abstract form of this form is doubly positive: $x \mid y$, where

form is neither x nor y but the distinction between x and y. This explicitly yields a self-referential form of form. "A difference-theoretical theory of form . . . treats forms as pure self-reference, made possible by the marking of the form as a boundary that separates two sides—made possible, in other words, by the fact that form is essentially a boundary."[48]

Two-sided form can be crossed over and back, once or repeatedly, without invoking the negation of form. One can still operate with one-sided concepts of form—form/substance, form/randomness, etc.—but these are now supplemented by the two-sided concept of form as the boundary of distinctions. With two-sided form, form reenters its own space. Two-sided form joins one-sided form with its shadow, nonform, constructing a second-order observation of the paradoxical form of classical ontological distinctions. Both are operationally valid but both have limits of application: both are bounded, not absolute. Observed from without—from another system—forms are virtually two-sided, but, from within, at the moment a system employs the differences they make they can be seen only on the side turned toward the inside of the system:

> The distinction of marked and unmarked is *one* distinction among others . . . , a frame that separates two sides and can be used to connect operations only at one side (at the positive side, at the inner side of the form) and not at the other side. The other side remains included, but as excluded. The excluded third, or the "interpretant" in the sense of Peirce, or the operation of observing in our theory, or the "parasite" in the sense of Michel Serres, or the "supplement" or "parergon" in Derrida's sense, is the active factor indeed, without which the world could not observe itself.[49]

When two-sided forms are personified, the metamorphs so created open up and dramatize the form of self-referential paradoxes. The unities created by holding the inside and the outside of distinctions together produce "mythological creatures" (see below), metamorphic chimeras fated to equivocate because the simultaneous observation of both sides of a distinction generates a paradox. But, as we witnessed in *A Midsummer Night's Dream*, metamorphic narratives construct cognitive spaces in which such paradoxical forms can take observable shape. David Wellbery notes how *"cultural systems tend to invent maverick terms that dramatize (or perform) the paradox of those operative terms and distinctions with which the culture conducts its observations. . . . [The] culture creates a kind of trickster, a unitary figure* in which both sides of a distinction coincide."[50] This is what occurs in tales of metamorphosis that splice human and nonhuman bodies and psyches together into composite creatures.

Two-Sided Forms in Beyond the Barrier

Two-sided forms inhabit Damon Knight's 1964 science fiction *Beyond the Barrier*, a futuristic fable that climaxes in an unforeseen moment of posthuman metamorphosis.[51] Here I will pass by, without interpretive remark, this novel's resonances with the contemporary work of Philip K. Dick and its pleasantly blunt allegories of class and race politics, to focus on the narrative's particular enactments of the paradoxes of form. In an amnesiac struggle to recover a buried past, the protagonist is accosted in the daylight by aliens masquerading as humans and at night by vivid dreams of possessing another identity in another world. The flip-flop of the storyworld back and forth from waking frame to dream frame is a case of vertical or ontological embedding—an oscillation of storyworlds—rapidly heightening the sense of general disorientation. Vertical embedding also emerges in a temporal mode as the main character is eventually transported from a late-twentieth century storyworld to a colony of humanity 20,000 years in the future.[52]

In my construction of the fabula, this future society is divided by species into the Lenlu Din, the Lenlu Om, and the Zugs. The Lenlu Om are green-skinned aliens brought to Earth to serve Man about 9,000 years prior (or 11,000 years in the future), but they also brought with them a virus that mutated upon arrival, wiping out all but a few immune humans. The Lenlu Din are the descendents of these survivors. They retain mastery over the Lenlu Om and have redeveloped advanced technologies including time travel. They are now an effete and bloated people ruled by the Highborn, a mad queen. They are also served by a lower caste of Lenlu Din, the Yani, who provide them with entertainment to while away their ultrarefined existence, and who also on occasion serve as guardians. For the Lenlu Din now live in fear of a newly arrived alien predator, the Zugs, who lay their eggs inside the bodies of the Lenlu Din. The Zugs, that is, practice a particularly stark and metamorphic form of reproductive embedding.

To fight the Zugs, the Lenlu Din created a mutant strain, a warrior class called the Shefthi. They have also built a New City and abandoned the old

Zug-infested one. But now they have come up with a final solution, a wall made of time, a Time Barrier that will exclude all beings but themselves from the future. The Barrier is a two-sided boundary conceived to enforce a one-sided form, for it is intended to produce a past–future dialectic that amounts to the antithesis Zugs / no Zugs. In addition, the Barrier will also snuff out the world lines of all the Lenlu Om. And, if the Barrier works as planned, the services of the powerful Shefthi will no longer be needed. As an extra measure of purification, to eliminate any threat the Shefthi might pose to the ruling caste of the Lenlu Din within the New City, prior to the completion of the Barrier they decide to eliminate all of them too, sending them on a bogus Zug-hunt that maroons them at random in the deep past. But, and this is the kicker, when the Lenlu Din switch the Barrier on, their sensors detect that *one* Zug has managed to stay alive and at large within the New City. Despite their desire for a one-sided solution, in operation the Barrier remains two-sided. It has not enforced the ontological solution that was to read: "the Lenlu Din exist, the Zugs do not." A frantic time-traveling S.O.S. is sent from the post-Barrier future to the pre-Barrier past, demanding that a Shefth be recovered and brought back to the future to hunt down and destroy this rogue Zug.

Present and past stand to each other as consciousness and memory, mental operations that may be distinguished by the "time barrier" that produces the markers of temporal difference attached to psychic constructions. Another two-sided form is presented by the simultaneously different personalities of a character in disguise. *Beyond the Barrier* begins with Douglas Naismith, a physics professor in the postnuclear Los Angeles of 1980, the seeming survivor of a military plane crash, who can remember only the four years of his life since, the first thirty-one a complete blank. By dividing the subject along the boundary between remembered and forgotten experiences, the state of amnesia is already an anticipation of a two-sided personality. Weird things begin happening to Naismith: a student accosts him with the incomprehensible question, "What is a Zug?" The student turns out to be a Lenlu Om in disguise, one of a couple that has come back in time in search of a Shefth, whom they intend to convert to *their* cause, the defense of the Lenlu Om against the impending holocaust of the Barrier. During this period they try to restore Naismith's Shefth identity. They program it on behalf of the Lenlu Om, then disguise it with a cover story designed to get him beyond the Barrier and through the Lenlu Din's sensors into the New City.

But, unbeknownst to Naismith or his Lenlu Om abductors, a Lenlu Din entertainer named Liss-Yani, who has also been sent in search of a Shefth, has been surveilling them and interfering in their efforts, sending Naismith dreams in which he appears as the entertainer Dar-Yani and experiences the conditions of life in the New City among the Lenlu Din. Eventually the Lenlu Om deduce that the Lenlu Din are already on to their scheme; thus Naismith is useless to them—they abandon him in a dead time zone, from which Liss-Yani rescues him. She notes in him some un-Shefthlike qualities—for instance, he doesn't kiss like a Shefth—but nevertheless they travel through spacetime from the dead zone to arrive at the City, actually an orbital space metropolis, at a moment in time just prior to the completion of the Barrier.

Once inside the City, Liss-Yani introduces Naismith to Prell, the director of the Barrier project, who allows him to view the construction of the Barrier through a portal. In this classic science-fiction moment, the entirety of the fantasy is exploited for its motivation of an extreme focalization. The vision of the narrative is taken into deep space and time as well as, here, into a deep level of matter under nanotechnological manipulation. No love-in-idleness is needed to make things change shape. And, instead of Oberon's high vantage overlooking the world below, we might think of this scene from *Beyond the Barrier* as presenting a *deep focalization*, in which the play of embedded levels involves the perceptual as well as the narrative agency. The narrative limns the cybernetic embeddedness of the psychic system within environments constituted by networks of social and technological systems:

> Suddenly there was a shimmer, a crackle, and a great round sheet of silvery reflection came into being. . . . As Naismith's vision adjusted to the scene, he began to make out serried ranks of dark objects, not visibly connected to one another, among which . . . human and robot forms came and went. . . .
>
> "This is the Barrier control network," the girl's voice explained. "They've been working on it for five years. It's almost finished. . . ."
>
> Part of the scene before them seemed to expand. Where one of the floating machines had been, there was a dim lattice of crystals, growing more shadowy and insubstantial as it swelled; then darkness; then a dazzle of faint prismatic light—tiny complexes in a vast three-dimensional array, growing steadily bigger . . .

Naismith caught his breath. He realized that he was seeing the very molecules that made up the substance of the machines that were being built in the next chamber. . . . The magnification increased. In luminous darkness, Naismith saw molecules scattered like tiny planets. A moving dot of light appeared, slowly traced a mathematical arc across the blackness. Other arcs of light sprang out from it, like ribs from a spinal column; slowly, the dots that were molecules drifted across to take position upon them. . . .

"It's beautiful," said Naismith.[53]

Naismith's response appears to be "brought into focus by a strict condition of perception," an aesthetic vision detached from the Barrier's practical purpose.[54] The problem is that a Shefth is not supposed to have aesthetic responses, and by his exclamation, Naismith has crossed a behavioral boundary that undoes his disguise. Prell begins to realize what Liss-Yani dimly surmised, that buried beneath Naismith is not the Shefth they take him to be. Somehow realizing Prell's realization, while still unaware of precisely what he realizes, Naismith strangles Prell to keep him from alerting anyone, then slips back through a portal to Liss-Yani. They proceed to the Highborn, who immediately questions Naismith's Shefth credentials. While he receives this audience with her, she receives a message from the future, repeating the alarming news that one Zug remains at large beyond the Barrier. Despite the simultaneous news that Prell is dead and Naismith is likely the one who killed him, it is decided to spare him long enough to test, here on the near temporal side before turning on the Time Barrier once and for all, whether he is in fact a Zug-killing Shefth. Naismith surprises himself and everyone else with an instinctive flash of martial reflexes leaving a remaining Zug captive dead on the floor. He keeps his life.

The next day the Barrier is activated: automatically all the Lenlu Om remaining in the New City collapse and die. Now beyond the Barrier, Naismith the reluctant Shefth is summoned once more to battle, but before he can go into action against the unknown solitary Zug, he too dies. That is, the human body masquerading as a Shefth that had called itself Naismith does. But the Zug entity embedded within that body now recovers its memory, for it is he and he is it. It had carried out its own anti-Barrier operation by implanting itself within a Shefth the Lenlu Din had tried to discard, and then doubling the disguise by substituting its Shefth host at the crash site for the dead body of the human Gordon Naismith, plucking his dogtags and assuming his human identity. Practical matters about the bodily forms in which the Zug entity performed these actions are left to the reader's imagination.

Breaking out of its human-body husk like an imago out of its chrysalis, the Zug now triumphs because there is no longer a Shefth anywhere to hunt it down. It takes control of the New City, and while having the Highborn prepared for dinner, contemplates what to do with the rest of the Lenlu Din, the pitiful remnant of future humanity. By rights it should wipe them out. But it finds itself "oppressed by the phantom personality of the man, Naismith," that is, by Gordon Naismith's human memories: "Every thought, every feeling that Naismith had had during the months their minds were linked together was recorded in his brain. It was not merely that he remembered Naismith: he *was* Naismith. He was a member of the race of conquerors; and he was also a man."[55] And so, in a fit of magnanimity, the Zug sets Liss-Yani and a male Entertainer free on the empty Earth, an Eve and Adam to propagate a new human evolution.

The narrative unfolds the enigmatic layers of a "mythological creature" constructed by the embedding of an alien seed (the Zug) within a posthuman mutant (the Shefth), then within a human being (Naismith). The two-sided protagonist of *Beyond the Barrier* develops from an amnesiac into a hybrid coupling of initially alien psychic systems, both of which remain intact in a higher-order being that has now internalized or fed back into itself its own species boundaries. Quickly the Zug/Man comes to appreciate its newly two-sided form: "How curious to think that this detached pleasure, half cool, half warm, was possible only to the mythological creature he had become."[56] The decisive revelation of posthuman metamorphosis is realized in the figural narrative situation of a mind beyond the human that yet retains its human traces. This form of posthuman framing by xenogenic biology will return in chapter 6 when we examine Octavia Butler's Xenogenesis trilogy. In both texts, themes of species extinction and continuance are folded into a form of metamorphic embedding, not of tales within tales, but of systems within systems. Reading the posthuman as a symbiosis of the human and the alien, embodied minds within social systems are fictively unraveled and reentered into bodies within bodies, and minds within minds.

FOUR

Metamorphosis and Embedding

Narrative Embedding and Cybernetic Form

Postmodern, self-reflexive texts as well as their cyberpunk spawn are well understood as fictive reworkings of cybernetic developments. Narratives of bodily transformation in the cybernetic era are often feedback loops reentering the forms of computation and communication technologies into the stories communicated. But those cultural thematics have left out a more intrinsic level of formal correspondence between recursive forms in the operation of systems and in the structures of narratives. Narrative embedding is the primary textual analog of systemic self-reference. The zigzag play and sequence of embedded and embedding narrative frames reenacts the essential paradoxicality in the operation of observing systems. Systems theory reconceives form not as the mere shape of things but as the cognitive boundaries by which thoughts or things are distinguished and observed. Embedded and embedding narrative frames assume precisely this self-referential form of form by marking the virtual edges of narrative structures.

For instance, the temporal unfolding of narrative observations needed to maintain the thread of a complex, deeply embedded diegesis involves re-membering or reconstituting each border crossing from one to another level of a storyworld, or from one storyworld to another.

American author John Barth has meditated both metafictively and criti-cally on the matter of embedded narration.[1] It is telling that one of the most profusely embedded narrative texts in recent memory, Barth's "Menelaiad" in *Lost in the Funhouse*, is also a postmodern adaptation of a famous meta-morphic episode from the *Odyssey*. In Book 4 of Homer's epic, Menelaus as character-narrator recounts for Telemachus how he had captured the sea god Proteus and held him fast despite the demigod's attempts to elude him by transforming rapidly into a series of slippery forms: water, fire, a lion, a serpent, a tree.[2] In Barth's tale, however, Menelaus narrates to an unmarked listener an account focused on the one being he could not hold onto: his wife Helen. Homer's Proteus combines with Barth's revision of Menelaus remembering his relationship with Helen to yield a story about the diffi-culty of holding onto elusive things. By transforming the Homeric Menel-aus's narration of the metamorphoses of Proteus into the postmodern Menelaus's rapid and multiple changes of narrative level, Barth puts the reader in the hero's sandals: to follow the story one must hold on fast while the shape-shifting of narrative voices carries one ever deeper into tales em-bedded within the frame tale. We learn to track our progress by counting the quotation marks that provide a running measure of the depth to which the narrator has embedded the story. Helen gets the last and deepest word, winged with eight levels (the outermost unmarked) of diegetic tension:

" ' " ' " ' " 'Love!' " ' " ' " ' "

In *Frameworks: Narrative Levels and Embedded Narrative*, William Nelles winds this thread back to the entrance of Barth's labyrinth of narrative embedding:

> The question "Why?" and the response "Love" are at this point octadiegetic narratives, on the eighth level of embedding: Helen's reply to Menelaus on their bridal bed (8th level) is quoted by Menelaus to Helen years later in Troy (7th level) within Menelaus's subsequent narration to Eidothea (6th level) which makes up part of Menelaus's narration to Proteus (5th level) which is set within Menelaus's narration to Helen on a ship after leaving Troy (4th level) as part of a story Menelaus narrates to Telemachus and Peisistratus in Sparta (3rd level) as part of a monologue between Menelaus and his own alter-ego (2nd level), the

entirety of which is embedded within the discourse of the voice of Menelaus-as-text narrating extradiegetically the entire story to the extradiegetic narratee.[3]

Barth's metafiction is an especially spectacular instance of the way that narrative embeddings put self-referential processes into literary operation. Embedded narrations elicit the observation of narrative observation by enacting the narrating of narration. Prompting recursive operations—the continuous reconnection of differential elements and levels—on the reader's part as well, stories within stories mimic the shifting of formal frames by which meanings telescope through levels of possible significations. Stories of metamorphosis simply raise these effects to another power: daemonic or transformative bodies, uncanny reframings of narrative identities, are further self-referential recursions of the formal structures of multiple and shifting narrative levels.

I. EMBEDDING AND LIFE

In his celebrated article "Partial Magic in the *Quixote*," Jorge Luis Borges observes how the virtual embeddedness of literary texts within the wider world invites exploitation by authors intent on playing with ontological distinctions: "Every novel is an ideal plane inserted into the realm of reality; Cervantes takes pleasure in confusing the objective and the subjective, the world of the reader and the world of the book."[4] Cervantes's narrator puts a twist on this virtual embeddedness of literary texts within the world at large by embedding the real existence of the first part of the *Quixote* within the fictive world constructed by its sequel. By placing a depiction of the text into the text itself, Cervantes turns the commonsense containment of books within the world into a paradoxical loop in which the reality of the book as a worldly object feeds back into its own fictive texture.

Genette terms narrative embedding *metadiegesis*, the emergence of additional narrating instances and hence narrative levels within storyworlds already being narrated. And once any metadiegetic level occurs in a text, narrative *metalepsis*—"intrusion by the extradiegetic narrator or narratee into the diegetic universe (or by diegetic characters into a metadiegetic universe, etc.), or the inverse . . . produc[ing] an effect of strangeness"[5]—is primed to follow as surely as a Möbius strip results from a simple 180-degree twist in the plane of a normal loop. We see metalepsis follow directly from metadiegesis when Borges points to the *Arabian Nights* as a prime example of this circular mode of narrative self-reference: "These inversions suggest that if the characters of a fictional work can be readers or spectators, we, its readers or spectators, can be fictitious."[6] Reviewing Borges's classic discussion under the term "self-embedding," Tzvetan Todorov notes that even without these explicitly metaleptic twists in the *Arabian Nights* and the *Quixote*, narrative embedding per se is already self-referential in its basic structure:

> Embedding is an articulation of the most essential property of all narrative. For the embedding narrative is the *narrative of a narrative*. By telling the story of another narrative, the first narrative achieves its fundamental theme and at the same time is reflected in this image of itself. The embedded narrative is the image of that great abstract narrative of which all the others are merely infinitesimal parts as well as the image of the embedding narrative which directly precedes it. To be the narrative of a narrative is the fate of all narrative which realizes itself through embedding.[7]

It is curious, however, that the topic of embedded narrative is linked by a number of authors, often in reference to the thematic repercussions of Scheherazade's embedded storytelling, to matters and/or metaphors of life and death. No metaphor is more dead than the vehicle "life" in phrases such as "the life of the mind" or, as in my concern here, "the life of the text" or "the life of the story." And yet the trope of narrative "vitality" animates some well-known and theoretically rigorous texts of structuralist narratology. One might expect that—given its formalism, demystifying intentions, and abstraction toward minimal units—structuralist discourse would be immune to such residual biomorphosis. Yet it often reinscribes it in the very act of its erasure.

We see this in Todorov's *The Poetics of Prose*. The chapter "Narrative-Men" begins with a definitive structuralist gesture—a discussion aimed at restricting psychological reification when considering literary characters. Thus Todorov distinguishes the "psychologism" of modern and modernist prose narratives from the "a-psychologism" of many epic and classical texts. This distinction marks the difference between "men" and "narrative-men," that is, between virtual persons and textual constructions.[8] Todorov notes that such "a-psychological narrative . . . is characterized by intransitive actions: action is important in itself and not as an indication of this or that character trait."[9] Put another way, when observed as a-psychological, narrative action is self-referential rather than the signifier of some character's virtual psychology. However, these matters of interpretive constructivism can cut in different directions. Once his discussion in this chapter shifts from characterization to narration per se, Todorov's insistence on the reflexivity of a-psychological narrative action triggers a reinscription of narrative "biologism," a kind of narrative vitalism adjacent to the psychological reification that he's just been arguing against.

The test case to which Todorov refers throughout "Narrative-Men" is the text of the *Arabian Nights*, which displays recursive or self-referential construction both in the a-psychologism of its narrative action and in the nested or embedded structure of Scheherazade's tales within tales. There is in this text a sort of implosion of form and theme: framed narrations call attention to narrating per se, and the burden of narrating is, for the character Scheherazade in particular, a matter of life or death. Todorov comments: "in the *Arabian Nights* . . . if all the characters incessantly tell stories, it is because this action has received a supreme consecration: narrating equals living. The most obvious example is that of Scheherazade herself, who lives exclusively to the degree that she can continue to tell stories; but this situation is ceaselessly repeated within the tale."[10] These ceaseless repetitions occur precisely within the tales embedded within the tale of Scheherazade. Within these tales within tales, "for the characters to be able to live, they must narrate. Thus the first narrative subdivides and multiplies into a thousand and one nights of narratives."[11]

In Todorov's discussion the single topic that elicits multiple inscriptions of narrative vitalism is embedded narration. This narrative structure is positively prolific in its ability to "subdivide and multiply," budding off new narratives like a fertilized ovum forming a multicellular zygote that produces an embryo that produces a fetus, etc. At the same time, this self-referential formative process of narrative gestation repeatedly underscores the crucially self-referential theme—"narrating equals living." After a brilliant rehearsal of the reciprocally supplementary relations of framed and framing narratives, Todorov gives the following summation:

> Such is the incessant proliferation of narratives in this marvelous story-machine, the *Arabian Nights*. Every narrative must make its own narration explicit; but to do so a new narrative must appear in which this narration is no more than a part of the story. Hence the narrating story always becomes a narrated story as well, in which the new story is reflected and finds its own image. Furthermore, every narrative must create new ones—within itself, in order that its characters can go

on living, and outside itself, so that the supplement it inevitably produces may be consumed there.[12]

On the topic of embedded narrative in the second edition of *Narratology*, Mieke Bal's discussion is infused with similar thematics of narrative vitalism. And this occurs with regard not only to *Arabian Nights*—"To the King and to Scheherazade narrating means life"—but also to modern examples of framed narration: "In Morrison's *Beloved*, a much more complex narrative, this principle—narration produces life—is also dramatized."[13] Refining the discussion of framed tales through her three-layer theory of narrative—text, story, fabula—Bal remarks that in *Beloved*, the embedded fabula accounts for both the existence and the shape of its frame: "the secondary narrators' joint efforts slowly narrate Beloved into life. . . . Narration is an act of creation. . . . The point of the narrative is, precisely, the creative power of story-telling itself, as a life-giving act."[14] Thus Bal rearticulates the link between textual embedding and narrative vitalism. A story within a story renders this link especially spectacular by arranging for its form—"story-telling itself"—and its theme—"a life-giving act"—to be cocreative. Bal concludes this section of her text by generalizing these dynamics to the reader as well, whose proper efforts of reception ensure that artworks are appreciated "not as a fixed collection of enshrined objects, but as an ongoing, live process. For some, even life-saving, for others just enlivening; for us all, part of life."[15]

Why is it that when two seminal narratologists in the structuralist tradition reach similar extremes of formalist inversion by describing the potentially infinite regress of narrative embedding, their discourses both flip over into a vitalist thematics? These conceptual tropes of narrative vitalism are directly connected, I submit, to the link between narrative embedding and the recursive operation of autopoietic systems. One key to the phenomenon we are trying to account for may be the pun embedded in the frame: the narrative frame is the body of the text at any given level, a potentially fertile body ready to reproduce and bud off new frames. In these strong figurations, narrative in metadiegetic dissemination comes alive, metamorphosing from the social circulation of communicated structures to the "living" progeny of reproduced narrating systems. We see that this mode of narrative circularity, the self-referentiality in the two-sided form of embedding and embedded narrative frames, is anything but a sterile formalism. Rather, we come, through the back door as it were, to the threshold of *autopoietic opera-tion* and an appreciation of the reentry of form into the form as a function that cuts across both biotic and metabiotic systems, and that rests ultimately on the necessarily mutual interembeddedness of abiotic, biotic, and metabi-otic systems.

Todorov and Bal skirt along without quite arriving at the formulation that the "life of narratives" and the continuation in being of their characters and storyworlds are actually predicated on processes that we can better un-derstand in terms of the autopoietic closure and contingencies of the subsys-tems provoked and sustained by the social circulation of literary narratives. In other words, the metabiotic, formal biomorphism of neocybernetic sys-tems theory, itself spun off of a theory of *biological autopoiesis*, explains the residual and compensatory vitalisms of structuralist narratology and recu-perates them for a systems-theoretical narratology.

2. BATESON'S PLAY FRAME

Another line of connection between narrative embedding and cybernetic form, William Nelles' *Frameworks* articulates an original series of perspec-tives on a main distinction among modes of narrative framing. He variously terms this distinction verbal and modal, horizontal and vertical, or epistemic and ontological embedding. Verbal embedding is horizontal and epistemic: different tellers' tales are embedded within a single overriding story frame, and the shifting among narrators concerns their differences as observers or epistemic agents, while more or less occupying the same level or realm of being. An example of horizontal embedding would be Mary Shelley's *Frankenstein*, where the monster's story is embedded within Victor's story, which is embedded within Walton's story. While stationed at different de-grees of embeddedness, each remains a part of the same early nineteenth-century storyworld of Europe and the British Isles—and thus at the same level of being within the main frame of the novel.[16]

Nelles distinguishes verbal, horizontal, or epistemological embedding from modal embedding, in which the shifting is vertical and ontological: here the same or different narrators are transported to and thus reframed within different storyworlds—for instance, modal borders are crossed in the transit "through the looking glass" from waking to dream worlds, from the present to the past or future, or from physical space to cyberspace. Verbal and modal forms of framing in no way exclude the other, however, and when combined, that complex or doubled narrative frame becomes, in Nelles's term, "'deep' . . . with both vertical and horizontal 'movement,' when the shift in narrator is accompanied by a shift in narrative level."[17] A classical example of a deep embedding framed by a *metamorphic* narrative would be the tale of Cupid and Psyche within Apuleius's *Golden Ass,* a framed story told within the metamorphic character-narrator Lucius's telling by the Bandit's Cook (horizontal embedding). But this shift of narrators transports the reader from the Hellenic story-present of Lucius the ass to the timeless world of the gods (vertical embedding), and embeds within Lucius's ludicrous character-bound narration the Bandits' Cook's high authorial or omniscient rendition of the perils of Psyche.[18]

Nelles continues: "This coincidence of differences at the boundary between embedded and embedding narratives is inevitably a site upon which to focus interpretation, automatically entailing additional structural and dramatic considerations that may provide sources of meaning."[19] But Nelles does not find any simple or necessary meaning informing the presence of narrative framing. Rather, the device is inherently ambivalent, capable of inducing unpredictable inversions of sense: "narrative embedding often has the paradoxical effect not only of producing the illusion of a more profound realism or aesthetic unity . . . but also of undercutting that illusion at the same time."[20] Indeed, if there is "any simple conclusion, it would be that of Gregory Bateson, that 'a frame is metacommunicative. Any message, which either explicitly or implicitly defines a frame, ipso facto gives the receiver instructions or aids in his attempt to understand the messages within the frame.'"[21]

Nelles's remark implies that narrative framing as a communicative device may be importantly related to the cybernetic discourse that informs Bateson's discussion, but he does not further unfold the matter. Rather, from Nelles's lack of contextualization (absence of framing) for the Bateson quote, one might assume that the latter's remarks were in fact referring to narrative embedding. But this topic is nearly—if not entirely—absent from the article in question, originally published in 1955, "A Theory of Play and Fantasy."[22] Instead, that article is primarily addressed to grasping the communicative mechanics at work in *play* altogether (of which narrative fiction would be a significant variety). As part of his larger campaign to redescribe the behavioral sciences through communications concepts, Bateson examines the "play frame" by employing a familiar repertoire of cybernetic ideas—self-reference, paradox, and feedback. Bateson notes that to succeed as a social behavior, play must be *framed as play*—that is, operationally bounded through communications among the players and set off as such from nonplay—and also, framed *in time*, again by means of signals that continuously convey the boundaries of play time to and maintain it for all involved in the game.

Bateson's pioneering cybernetic discussion resonates with Nelles' narratology of embedding through shared metaphors of frames and also of *levels*—for Bateson, levels of communication. Bateson explains how various *meta*levels frame and so transform the content of discourse at other communicative levels: "Human verbal communication can operate and always does operate at many contrasting levels of abstraction. These range in two directions from the seemingly simple denotative level."[23] At the denotative level, the default mode of normal speech, the topic or "subject" of discourse is taken to be some objective thing or event in the world. But "one range or set of these more abstract levels includes those explicit or implicit messages where the subject of discourse is the language. We will call these metalinguistic."[24] At the metalinguistic level, the attention of the participants in a dialogue shifts to the shared code by which they refer to the world, a shift often performed to ensure semantic redundancy, the mutual awareness that interlocutors understand the same things by the same words. Bateson continues, "the other set of levels of abstraction we will call metacommunicative. . . . In these, the subject of discourse is the relationship between the speakers."[25] At the metacommunicative level, the attention of the speakers shifts from the heteroreferential objects and code-referential tokens of their discussion to the self-referential matter of their own behavioral motives for communication, prior to any questions of objective or linguistic reference. For instance, metacommunications try to ensure that transitions from serious to playful, from earnest to ironic discourse, and back again, are understood by all listeners. Bateson's metalinguistic and metacommunicative frames clearly anticipate our neocybernetic connections to the social and psychic domains of narrative theory.

Fascinatingly, Bateson relates how these communicative considerations were crystallized for him not by incidents of human interplay, spoken or otherwise, but by frolicking monkeys at a San Francisco zoo. He observed

how some of their social play was to feign aggression by delivering each
other bites that were not really bites, but "nips." That is, they used the nip
to frame the bite (the bite that doesn't come, except as a nip), to indicate
through this gestural metacommunication (neocybernetics' "formal distinction," Spencer-Brown's "cancellation") a playful reframing of a behavior
that would otherwise be received without pleasure. The nip was at once a
transitive communication—the monkeys' way of playing—and an intransitive metacommunication, testing moment by moment others' understanding that they were not at present in earnest. The brilliance of Bateson's
analysis emerges when he seizes this monkey business—his ethological observation of the play frame in simian action—as parallel to human communicative behaviors containing

> those elements which necessarily generate a paradox of the Russellian or Epi-
> menides type—a negative statement containing an implicit negative metastate-
> ment. Expanded, the statement "this is play" looks something like this: "These
> actions in which we now engage do not denote what those actions *for which they
> stand* would denote." . . . The playful nip denotes the bite, but it does not denote
> what would be denoted by the bite.[26]

Put another way, Bateson's play frame is a mode of metaphorical bracketing in behavioral action. The play frame suspends and transforms denotation, constructing a figuration of the literal, contextualizing semiotic
elements to transform their semantics. Numerous life forms other than humans are capable of such denotation and its bracketing, generating a bonus
of pleasure that makes social play a desirable activity. As such, the phenomenon of play implicitly breaks down overly rigid distinctions between human
and animal social systems and cognitive capacities. This was a message always ready to be taken from the long tradition of metamorphic narratives.
This blurring of ontological boundaries among living beings is also of a
piece with the posthuman trend of cybernetics in general toward rapproche-
ment between humans and nonhumans.

The play frame links narrative play to philosophical discussions of paradox. As a logical form, paradox results from purported deviations in proper
relations among classes or levels of verbal statements. In the article at hand,
Bateson goes on to remark: "What has previously been said about play can
be used as an introductory example for the discussion of frames and contexts. In sum, it is our hypothesis that the message 'This is play' establishes

a paradoxical frame comparable to Epimenides' paradox" (Bateson, 184). In this classical example of verbal self-contradiction, the paradox is generated by the self-reference: Epimenides the Cretan virtually includes himself in the set of liars predicated by his universal premise, "all Cretans are liars." The cyberneticist would say that he feeds back his own output as input, as the major proposition of his enthymeme circles back on the unstated minor—his own status as a member of the set of Cretans. That is, he places himself both outside and inside the conceptual frame (the logical set) his proposition constructs. As a result, if he's telling the truth he's lying and if he's lying he's telling the truth. One is left with a both/and situation (a double positivity) that cannot be resolved as either one or the other. The paradox generates a cognitive flip-flop that suspends one's normal expectations about denotative statements. Rather, one is precipitated into an oscillation comparable to the metaleptic self-embedding of a narrative into its own narration: the embedded story (e.g., part one of *Quixote*) both is and is not the embedding story (e.g., part two of *Quixote*). Similarly, to summarize Bateson's exposition, the message "this is play" implies that x, once it is placed and as long as it is maintained within the play frame, both is and is not x: the nip both denotes and does not denote the bite.[27] Perhaps this would be better stated as: x is simultaneously both x and y. The nip is both a nip and a bite, but a nip-bite rather than a bite-bite.

Bateson's implicit overlaps among philosophical, communicative, psychological, and narrative contexts allow Nelles to extract a theoretical summary for his own discussion of narrative embedding. To repeat Nelles's citation of Bateson's conclusion about the play frame: "the frame is metacommunicative. Any message, which either explicitly or implicitly defines a frame, *ipso facto* gives the receiver instructions or aids in his attempt to understand the messages within the frame."[28] But Nelles's use of this particular formulation of metacommunicative frames in relation to narrative embedding obscures somewhat the more playful ramifications of Bateson's wider discussion: the propensity or potential of all such frames to baffle as well as assist the receiver of the message, to precipitate their observer into the play of paradox rather than carry them to any one haven of interpretive clarity. Indeed, the flip-flop effect of narrative framing, as powerfully propounded by Todorov—its constitution of an inherently paradoxical boundary—draws it most powerfully into the neocybernetic discussion of forms and operational boundaries.

Todorov's "Narrative-Men" points in this direction when his discussion takes the following crucial turn: "Let us attempt to take the opposite point of view, no longer that of the embedding narrative but that of the embedded narrative, and inquire: why does the embedded narrative need to be included within another narrative? How can we account for the fact that it is not self-sufficient but requires an extension, a context in which it becomes simply a part of another narrative?"[29] Todorov cannily flips his discussion of narrative framing inside out, demonstrating in the process that such an inversion is always ready to be called forth.

From looking down into the abyss of embeddedness, as it were, Todorov has the abyss look back out at embeddingness. No matter which way one turns, there is no way out of the underlying paradox that there is no final frame at which to halt the play of embedding. Any narrative is virtually abyssal, part of an infinitely unsupported circle. No frame stands alone or on its own. Each relies upon of some further, inner or outer frame to prop it up internally or externally, which circumstance necessarily applies as well to that further frame. Todorov continues:

> If we consider the narrative in this way, not as enclosing other narratives but as being enclosed by them, a curious property is revealed. Each narrative seems to have something excessive, a supplement which remains outside the closed form produced by the development of the plot. At the same time, and for this very reason, this something-more, proper to the narrative, is also something-less. The supplement is also a lack; in order to supply this lack created by the supplement, another narrative is necessary. . . . There is no reason for this process to stop anywhere. The attempt to supply is therefore vain—there will always be a supplement awaiting a narrative-to-come.[30]

Italo Calvino also limns this regress of supplementarity in "Levels of Reality in Literature" when he speaks of an "invisible colon" that hovers over every narrative frame, no matter how extradiegetic, marking a boundary that virtually inscribes it within some further narrative or discursive context. "This colon is a very important articulated 'joint,'" says Calvino, "and I would call it the headstone of narrative at all times in all lands. Not only because one of the most widespread structures of written narrative has always been that of stories inserted into another story that acts as a frame, but also because where the frame does not exist we may infer an invisible colon that starts off the discourse and introduces the whole work."[31]

Todorov's evocation of supplement and lack are contemporary with Derrida's treatments of the *supplement*. His contemporaneous discussion of the *parergon*—that which is beside (*para-*) the work (*ergon*)—indicates the general supplementarity of frames in line with Todorov's particular description of narrative frames: "What constitutes . . . *parerga* is not simply their exteriority as a surplus, it is the internal structural link which rivets them to the lack in the interiority of the *ergon*."[32] It also shows that frames work this way irrespective of their intra- or extratextuality. That is, Derrida's *parergon* is a functional generalization of the notion of the *pharmakon*—his metaphor for writing as an ambivalent external–internal elixir, one that both poisons and props up the mind—in that literal or material frames—for example, picture frames or sculptural pedestals—operate to precisely the same supplemental effect as textual or conceptual frames.[33]

Derrida's *parergon* functions not only at the boundary between different narrations, as with Todorov's *invertible frame* and Calvino's *invisible colon*; not only between works of art and their physical and historical surroundings; but also at the boundaries that systems construct between conceptual and empirical, imaginary and real, self- and heteroreferential observations. Classical ontology is defrocked by the knowledge that there is no region of being left intact by the operations of framing. As far as psychic systems are capable of marking them, the elements and relations of the world in its complexity are always already abyssal. Nothing can come forth to us without a frame that both cuts it out from the unmarked state and renders it contingent upon the supplement constituted by that act of distinction. Within this universal frame, narrative embedding is implicitly a way of reframing the ontological ramifications of these epistemological circumstances and precipitating them into the paradoxes of literary play.

The "partial magic" of literary self-reference conjures up a truer image of knowledge—underscoring its necessary constructedness—than that offered by the linear denotations of literary realism. There is also something partially magical about the parergon as frame-in-general. One can both pull it out of the hat like Calvino's colon and presto! make it disappear. Its double-sidedness renders it self-absconding. Derrida describes the effect:

> The parergonal frame stands out against two grounds [*fonds*], but with respect to each of these two grounds, it merges [*se fond*] into the other. With respect to the

work which can serve as a ground for it, it merges into the wall, and then, gradually, into the general text. With respect to the background which the general text is, it merges into the work which stands out against the general background. There is always a form on a ground, but the *parergon* is a form which has as its traditional determination not that it stands out but that it disappears, buries itself, effaces itself, melts away at the moment it deploys its greatest energy.[34]

Compare this to George Spencer-Brown's *form of cancellation* (figure 3.3). In the vocabulary of *Laws of Form*, this notation reads: the recrossing of a distinction from the marked to the unmarked side suspends the indication of the distinction. But if read as an icon or rebus of the parergon, positioned between the work and the world, the equation given by the form of cancellation goes: *the frame both appears and disappears.* The deconstructive construction of the frame as parergon rhymes precisely with the epistemological constructivism of neocybernetic systems theory.[35] The difference would be that Todorov and Bal's residually structuralist discourses do not yet lend their frame-narrative structures the temporal and operational character that systems theory supplies and emphasizes. Systems theory supplements this lack in structuralism by reframing it, embedding the story of structures within the narrative of systems.

The Cyberiad

I. PERFECTION

First published in Polish in 1967, coming into English translation in 1974, Stanislaw Lem's *The Cyberiad: Fables for the Cybernetic Age* is a complex satire of overreaching among scientific fabricators and political engineers.[36] The narrative is told by an unnamed, far-future chronicler looking back at the epic adventures of posthuman heroes Trurl and Klapaucius. Weaving together medieval legend and futuristic comedy, Lem sets forth a pair of technological quasi-objects that troll the galaxy for booty and glory. Trurl and Klapaucius are robots, machine beings who build machines. While hiring themselves out to the highest bidder to design and produce military weaponry or other intelligent machinery, these cybernetic constructors also encounter a series of scholars and seers, whose tales elicit covert critiques of scientistic constructions—the misapplication of scientific ideas to social and political schemes—and of their philosophical justifications.

Metamorphic stories are likely vehicles for political allegory. Tropes of bodily metamorphosis—its evolutionary histories and its existential limitations—are important vehicles for the technopolitical allegory here. When the narrative becomes most self-referential about its own metamorphic devices, it also delves the deepest into the epistemological and ontological repercussions of cybernetic ideas. The key term critically scrutinized by the text is *perfection*. It is a term going back to classical philosophy and theology—the eternal perfection of the ideal or the divine as both the source and the goal of individual and social behavior. In this context, change or metamorphosis is either the corruption of perfection or the process of returning once more to the perfect unity of the original essence. Lem's figures play on this long doctrinal background while they update the notions of social perfection and personal or bodily metamorphosis by reference to scientific discourses of evolution and cybernetics.

The critique of "perfection" in *The Cyberiad* is first made explicit in the title of "The Seventh Sally, or How Trurl's Own Perfection Led to No Good." When a simulated social world in a box of his construction veers off on an unforeseen political evolution, he declares to Klapaucius, "my purpose was simply to fashion a simulator of statehood, a model cybernetically perfect, nothing more!"[37] It is models, then, not the things modeled, that are the objects of cybernetic perfection. But the perfectionism of first-order cybernetics in its positivistic mode—its drive for ever more "perfect" mechanical simulation of physiological and mental functions—is easily displaced from scenes of technical construction and virtual reification to real projects, such as Dr. Moreau's, for the technological "improvement" of individuals and societies.

Much of the cybernetic detail of the novel, then, is an allegorical cloak for social commentary. But the social also performs allegorical duty for the cybernetic content. Both Cold-War collectivism and first-order cybernetics are exposed in their complicities with forms of systematic coercion. The deeper satires of *The Cyberiad* are aimed at modern science in its most technoid manifestations: its collusions with control schemes for bodily and social perfection. Given that Norbert Wiener coined the name *cybernetics*, derived from the Greek for "steersman," to refer to the governance of systems, this science is already embedded with political connotations tailor-made for Lem's satirical role. The novel capitalizes on cybernetics' first incarnation—first-order cybernetics as compounded of the sciences of com-putation, information theory, control theory, and artificial intelligence—mechanical–electronic calculation, duplication, replication, simulation, symbolic modeling, especially the technological re-creation of biological organs (brains) and organic beings (persons). The novel's characters are parodic signifiers of the heroic constructivism in cybernetic engineering culture: the mid-century technoscientific culture centered on the invention of digital computing itself.

Trurl and his friendly rival Klapaucius are comic allusions to John von Neumann's 1948 Hixon lecture on the logical complexity of electronic computation, "The General and Logical Theory of Automata," his seminal discussion of the mechanical simulation of organic reproduction. The figure of the *constructor* comes from the end of this article, in a section about *"Automata That Produce Automata."* The main protagonist of *The Cyberiad* (the constructor Trurl) is literally "an automaton whose output is other autom-ata."[38] The topic of computation turns self-referential on this matter of a system's capacity for self-reproduction, when von Neumann envisions a scene of digital origin, a logical scenario for Turing machines to reproduce themselves, involving the replication of a set of instructions to guide the assembly, from suitable elements or "elementary parts," of a separate but identical offspring:

> The problem of self-reproduction can then be stated like this: Can one build an aggregate out of such elements in such a manner that if it is put into a reservoir, in which there float all these elements in large numbers, it will then begin to construct other aggregates, each of which will at the end turn out to be another automaton exactly like the original one? This is feasible. . . . The constructing automaton is supposed to be placed in a reservoir in which all elementary compo-nents in large numbers are floating, and it will effect its construction in that milieu.[39]

The mythography of this scene could be construed as a grotto or subterra-nean pool, an amniotic medium within a womb-like space for the parent machine's physical reconstruction of "perfectly" identical mechanical off-spring. Lem's narrative runs on the notion that machines will evolve such organs, that one day intelligent machines will leave us behind to perfect their own progeny. This is one future mythology of the posthuman. But as *The Cyberiad* unfolds, we find that there are many such mythologies—an origin myth for every occasion.

The long tradition of metamorphic fables often plays on the classical separation of body and soul, matter and form—an understanding of form and
structure from which the concept of system is absent. As a narrative device,
metamorphosis can either favor a mutual interplay of medium and form, or
subordinate the accident of the body to the essence of the soul. When first-
order cybernetics valorizes informatic controls over material contingencies,
input–output performance over the materiality of the systems doing the
performing, it lends itself to the latter mode of hierarchical dualism. Control theory then joins Neoplatonic allegory in rendering the body as a "trivial machine"—the material container for an immaterial agency—which
formal principle is taken to be able to exist either in attachment to or in
detachment from a worldly receptacle that, for its own part, cannot stay in
existence in detachment from that immaterial essence or form. Metamorphosis on this model is not the transformation but merely the substitution
of bodies through the transposition of minds conceived as independent of
bodily conditionings. "The Fifth Sally or the Mischief of King Balerion"
uses such first-cybernetic dualism to construct a comical duel—a tour de
force in this dualist idiom of classical metamorphosis.

Metamorphosis enters the tale of King Balerion through the game of
hide-and-seek. From the heifer in the myth of Io and the ass in *The Golden
Ass* to Gregor in Kafka's *Metamorphosis* and the aptly named Mr. Hyde, it is
proverbial that the metamorphic body is a place of concealment: an inadvertent or deliberate masking of identity. Especially when the metamorphic
body lacks human voice, the identity of the metamorphic agent is hidden,
or stranded, within it. Trurl and Klapaucius enter the scene when Balerion
announces a prize for "the best hiding place in all the world."[40] Anxious for
a sale, they peddle him a "portable bilateral personality transformer, with
retroversible feedback, of course. Using it, any two individuals could quickly
exchange minds. The device, fitted onto one's head, resembled a pair of
horns . . . . Through one horn, one's own psyche flowed into the other,
and through the other, the other into one's own."[41] Trurl demonstrates for
Balerion how to fit the horns onto one's head. The tale is jump-started
when Balerion puts this chiastic personality transfer into effect, butts into
Trurl, and, in the body of Trurl, with horns attached, runs off for the game
of his life.

As implied by the inconsequence of the given body in a first-cybernetic regime of disembodied information, here is a daemonic tale of cybermorphosis—complete with "devil's horns." Lem also taps into the classical Hermetic complex bound up in metamorphic stories, the mythic signature of Hermes the thief on the daemonic trades that lead to the transfer of title to, or theft of, another's body or soul. But let us stay with the first-order cybernetic analogs. The first topic in part one of Ashby's *Introduction to Cybernetics* is change—in particular, not continuous but digital, state-to-state changes called *transitions*. When one has a system of interlocking transitions, "such a set of transitions, on a set of operands, is a *transformation*."[42] An example "is given by the simple coding that turns each letter of a message to the one that follows in the alphabet":

$A \rightarrow B$
$B \rightarrow C$
. . .
$Y \rightarrow Z$
$Z \rightarrow A$[43]

"The Mischief of King Balerion" pursues a slightly more complicated set of transformations in a way that, formally considered, strips the fabula of this narrative down to the iterations of an algorithm:

> Start: The mind of Balerion takes over the body of Trurl; the mind of Trurl enters the body of Balerion.
> → The mind of Balerion takes over the body of the Sailor; the mind of the Sailor enters the body of Trurl.
> → The mind of Balerion takes over the body of the Police Commissioner; the mind of the Commissioner enters the body of the Sailor.

Clearly this sequence could continue forever, as long as Balerion repossesses the transforming horns after each bodily switch, that is, as long as the technology of metamorphosis is transferred along with every transposition of the agent willing the transformations. If these conditions remain unaltered, the form of the set of transformations will remain transitive, and Balerion will remain hidden forever in new bodies, leaving a trail of victims whose bodies and minds no longer match.

Further discussing cybernetic transformation, Ashby addresses the notion of *closure*: "When an operator acts on a set of operands it may happen that the set of transforms obtained contains no element that is not already present in the set of operands, i.e., the transformation creates no new element [as in the simple alphabetical coding above]. . . . When this occurs, the set of operands is *closed* under the transformation. . . . [But] if either is altered the closure may alter."[44] The fabula of "The Mischief of King Balerion" is resolved when the closure is opened enough to alter to a different closure: Trurl's colleague Klapaucius opens the set of transformations by introducing himself into it as an operator as well as an operand. Thus, at a strategic moment, he hoodwinks Balerion (currently in the body of the Commissioner) into momentarily relinquishing the horns, steals them back, puts them on, and butts Balerion the way Balerion had initially butted Trurl. This reversal of agency—dispossessing Balerion of the horns—sets into motion a second cascade of transformations: some restore prior relations, others institute new ones. The key is that certain bodies in Balerion's kingdom command authority, irrespective of the minds they contain.

→ The mind of Klapaucius takes over the body of the Commissioner; the mind of Balerion enters the body of Klapaucius.

Klapaucius uses the authority of his new body as well as his control of the horns to steer the metamorphic deformations of the social system to a new equilibrium:

→ The mind of Klapaucius retakes the body of Klapaucius; the mind of Balerion is returned to the body of the Commissioner.
→ The mind of Trurl is returned to the body of Trurl; the mind of the Sailor is given the body of Balerion.
→ The mind of Balerion enters the body of the cuckoo clock; the mind of the cuckoo clock enters the body of the Commissioner.

Thus all ends happily in the former kingdom of Balerion, as Trurl gets his body back; the mind of the Sailor—the ideal cybernetic steersman—now holds sway in the King's body; the police commission is run just as efficiently by a cuckoo clock; and for good measure, the mind of the Commissioner is left stranded for good in the body of the Sailor, which is imprisoned in the Commissioner's own jail.

In precybernetic metamorphoses, where the story involves a reversal of, say, human and animal bodies (generally without reference to the animal mind), comic endings typically feature a "proper" reunion of (human) body and soul: both Ovid's Io and Apuleius's Lucius find salvation from transformation by return to their proper bodies. Conversely, Dr. Jekyll's tragedy is finally no longer able to slough off his Hyde-out. In this tale Lem plays out mock-epic metamorphoses as a satire on the first-order cybernetic doctrines that analogize the informatic structures of computer programs to the detachable psyches of classical metaphysics. The informational transfer of the soul or mind or psyche from one mode of material embodiment to another becomes a mainstay of cybernetic science fantasy—cyberpunk in particular—generally without the satirical pressure Lem placed on it in the mid 60s.

Caught up in his tale of Balerion, however, we are likely to forget that the characters whose minds and bodies are involved in the episode *are all robots*, machine beings for which there is presumably no "duality" of mind and body—both are "mechanical." Thus the tale also satirizes its modern readers' willingness to override the organic–mechanical distinction while indulging in the hoariest highjinks of the mind–body split. Lem's tale exposes the ease with which theological or philosophical dualisms can transform mainstream cybernetics into a nostalgic idealism playing over an inert body. Construed as a pattern, not a thing, an informatic psyche may be inherently transportable and so detachable from any "original" bodily instantiation, but only as long as some other material carrier is available to bear the inscription of that pattern—metempsychosis rather than disembodiment altogether.

The matter of networks of recursive couplings, in which systems are grasped as semiautonomous yet codependent upon their environments and on other systems within their environments, is already there in Ashby.[45] Second-order cybernetics moves past these first-cybernetic equivocations to focus on the feedback of medium into form and the transformation of forms into higher-order media for metaforms. Inspired by the work of Ashby's colleague Heinz von Foerster, the neocybernetic turn of the 1970s develops concepts of cybernetic form that include rather than exclude their material environments. Lem writes these cybernetic fictions concurrently with von Foerster's transition from the first-order self-organization of structures to the second-order observation of observing systems. From literary and scientific angles, Lem and von Foerster put the paradoxes of form into serious play in discourses emphasizing the ways that the noise of the real—the embodiment of systems, the uncertain complexities of environments—complicates the orders inscribed in the formal informatics of systems.

Framed narratives suggest the embedding of worlds within worlds, such as that of the Microminians, the overly-perfected world inside a box inside the storyworld of Trurl's Seventh Sally. Stories within stories foster the sense of narrative embodiment within a textual environment: their storyworlds are that much more obviously contained within wider contexts, and whatever sense such stories may make is not "independent of the medium"—the frame narrative(s) in which they occur. *The Cyberiad*'s explicit deployment of cybernetic themes elicits the formal resonances between cybernetic and narrative self-reference. Especially in the form of embedded narratives, the novel weaves itself out of self-referential constructions, produces fictive scenarios out of the feedback of cybernetic content and cybernetic form.

After the Seven Sallies of Trurl and Klapaucius, *The Cyberiad* shifts gears with a seventy-page cascade of interlocking, multiply embedded stories: the "Tale of the Three Storytelling Machines of King Genius." The bard tells how Trurl built the King three machines, media devices that, when turned on, proceed to narrate the further adventures of Trurl, at times in the voice of Trurl. With Trurl and the King joining the reader as audience, the narration of the second storytelling machine begins: "Here is a story, a nest of stories . . . about Trurl the constructor."[46] The storytelling machines being told by the external narrator (a storytelling machine?) are now both inside and outside the stories telling of Trurl the constructor. Here is a two-sided narrative form—a remediation of remediation—with a vengeance.

After the first storytelling machine within the first-degree story comports itself well, the narrative comes back to the King Genius frame for some significant digressions on self- and narrative transformation. Before proceeding further with the interplay of stories on multiple levels, Lem's narration pauses for a moment of narrative Zen, as Trurl pulls the bottom out from under his own storyworld by self-reflection on his status as a character on a particular level of a multileveled metafiction. He follows that with another comment connecting the perfection theme to narratives of evolutionary and individual metamorphosis. Wedding metafictive form and the metafictive narratee's reading experience to the mobile constructivism of neocybernetic cognition, Trurl's commentary sets the play of two-sided forms—the procession of reframings in both narrative structures and system operations—against the image of perfection in the semantic content of the tales. Trurl asks the King:

"Suppose that which is taking place here and now is not reality, but only a tale, a tale of some higher order that contains within it the tale of the machine: a reader might well wonder why you and your companions are shaped like spheres, inasmuch as that sphericality serves no purpose in the narration."[47]

The narrative dilates at this moment to take in a speech we will examine in detail in the next section, King Genius's technoevolutionary chronicle of the machinic obliteration of the bodily appearance of "wet and spongy beings" by cybernetic self-transformation into perfect spheres. Trurl rejoins with an ethical proposition about the engineering of perfection into "bodies":

It is always best when an intelligent being cannot alter its own form, for such freedom is truly a torment. He who must be what he is, may curse his fate, but cannot change it; on the other hand, he who can transform himself has no one but himself to blame for his failings, no one but himself to hold responsible for his dissatisfactions. However, I did not come here, O King, to give you a lecture on the General Theory of Self-Construction.[48]

Beyond the obvious political moral and the allusion to von Neumann's "General and Logical Theory of Automata," I would take the author's point in the same key as the proto-neocybernetic reply to the unbounded play of identity lampooned by the tale of King Balerion in the Fifth Sally. One cannot simply will the alteration of or wish away the formal contingencies of self-referential systems: for instance, without functioning operational boundaries, autopoiesis in the cell or the psyche is not going to happen. Information leads to ordering as much from the binding of forms within media as from the transmission or replication of those forms. Self-construction—auto-poiesis—concerns the maintenance of form through the processing and binding (to memory or forgetting) of differences. What is properly and constantly transformed are the *relations* of any given form to other levels of observation, through systemic devices that shift narrative or other cognitive frames. Again, what is properly and constantly transformed are the relations of any given system to its environment and to the other systems in that environment—the forms of the couplings and embeddings that allow for the emergence of higher-order associations.

In the King Genius tales this operational dynamic of relational transposition is simulated by multiple shifts from one to another storytelling machine, and from one to another story by the second and third machines.

Flipped on, the second storytelling machine retails episodes of literal tyranny and symbolic vengeance embedded to the second, third, and fourth degrees. For instance, for "King Thumbscrew the Third, ruler of Tyrannia, who wished to learn the means of achieving perfection," the Trurl avatar narrated by Trurl's storytelling machine in the King Genius frame offers an account of his witness of the flogging by Vendetius Ultor of Amentia of Malaputz *vel* Malapusticus Pandemonius. Embedded to the third degree, Vendetius tells the story of the evolution of Automatus Sapiens. But that lesson in the evolutionary history of the *Cyberiad* storyworld is primarily prologue for a lament about more recent social developments.

It turns out that Vendetius and his cohorts are flogging the cybernetic effigy of Pandemonius, "an infernal intellectrician and efficiency expert" whose schemes for social engineering wreaked only "chaos and carnage, and a devastating decline in the vital voltage, an enervated emf and energy dissipation everywhere."[49] Lem embeds his skepticism about the collusions of technoscientism and socialism in the most deeply embedded story in the novel so far. As is common with politically coded satire under threat of censorship, he buffers his most heretical critiques of the Eastern Bloc under the most layers of symbolic mediation. The model for the efficiency expert in question is Frederick Winslow Taylor, and this spoof of Soviet Taylorization is the most direct attack on the Soviet state in the novel.[50] Lem renders an Eastern-Bloc view of the troubling ease of alliance between perfectionistic technosciences and perfectionistic state apparatuses. Overreliance on scientistic predictions breeds cynicism, inducing the sorts of inertia that break social systems down.

The final sequence given by the second storytelling machine pushes the device of embedded narrative beyond its already salient role in the text, imploding it through a runaway positive-feedback loop. This apocalypse of self-reference is centered on King Zipperupus, a self-absorbed erotomaniac plotted against by his vizier Subtillion, "an adept in mental engineering," who proposes to seize power by imprisoning the King in the folds of his own fantasy. He baits a trap with narratives of lust, " 'which he will seize and, in so seizing, of his own volition plunge into figments and mad fictions, sink into dreams lurking within dreams. . . . He'll never get back to reality alive!' "[51] That is to say, Subtillion plots to derange the mind of King Zipperupus by embedding it into embedded narratives, luring it into irreversible immersion within the virtual storyworlds of media devices. In an astonishing tour de force, working in the mid 1960s merely from the internal resonance of narrative and cybernetic forms, Lem lays out the lure and the architecture of a technology of virtual reality that will only take on real narrative and technological substance in the mid 1980s.[52]

Subtillion presents Zipperupus with three machines that produce worlds in a box. Anticipating virtual reality engines and recalling fairytale motifs of the three caskets, the three dream cabinets are also replicants of the three storytelling machines lying at a prior level of the narrative, the second of which is currently operating in the background to produce the narration now unfolding. But now, the oral storytelling machine in the King Genius frame is narrating an embedded story about a different sort of storytelling machine, one that works by the immersion of its observer into a simulated multisensory environment. Entering and exiting Subtillion's pornographic dream-traps, at first Zipperupus maintains his hold on—retains control of— his self-projection from one narrative level to another. For several adventures—"Alacritus the Knight and Fair Ramolda, Daughter of Heteronius," "The Marvellous Mattress of Princess Bounce," "Bliss in the Eightfold Embrace of Octopauline"—just when the stories corner him in desperate situations, the King's own cowardice or cupidity rescue him from the brink of destruction.

Zipperupus assumes the position of the reader of a deeply embedded metafiction. Even though the "solid ground" of any storyworld is a mirage or arbitrary epistemological construction, on the receiving end the gambit is always to hold onto "reality"—the initial reference frame. No matter how deep one goes, the reference frame orienting all the other narrative frames at play must be recoverable if one is not to lose their moorings completely. Subtillion anticipates, however, that if he can keep the game in play, the nature of the dreams will eventually slip into a blind spot and close up, luring the king into an infinite regress of indistinguishable levels of embedding. A vertiginous view of those levels appears in the "Wockle Weed" dream, when a witch's brew is splashed and leaves "a black hole gaping in the floor, and through it one could see another hole, a hole in the dream itself . . . clear through to reality," where Subtillion is hopping on a burnt foot.[53] Incorrigible, Zipperupus chooses next to visit "The Wedding Night of Princess Ineffabelle." Within this world he reads in an old volume about the beauty of the Princess, then sets off on a quest for her, only to discover that she has been dead for five hundred years. "An individual of patriarchal appearance" offers him a view of "a fair informational facsimile . . . all in yon Black Box."[54] Ravished by this vision within a vision, the King tries to climb into the box:

> "Madman! Wouldst thou attempt the impossible?! For no being made of matter can ever enter a system that is naught but the flux and swirl of alphanumerical elements, discontinuous integer configurations, the abstract stuff of digits!"
>
> "But I must, I must!" bellowed Zipperupus. . . . The old sage then said:
>
> "If such is thy inalterable desire, there *is* a way I can connect thee to the Princess Ineffabelle, but first thou must part with thy present form, for I shall take thy appurtenant coordinates and make a program of thee, atom by atom, and place thy simulation in that world."[55]

Agreeing to submit to a "test run," Zipperupus peers into the repro-grammed Black Box "and saw himself sitting by the fire and reading in an ancient book. . . . And so on."[56] Turning away from his own mise en abyme, again Zipperupus makes ready to flee from the ontology trap set for him, when "at that very moment the dream burst like a bubble beneath his feet" and he lands once again at the feet of Subtillion, who seizes on his disorien-tation to foist him one last time into a virtual limbo of no return, "Mona Lisa, or the Labyrinth of Sweet Infinity": "And he pushed in the plug and entered the dream, only to find himself still himself, Zipperupus, standing in the palace vestibule, and at his side, Subtillion."[57]

The narration shifts to present tense to capture Zipperupus's ultimate capture within Subtillion's virtual labyrinth:

> Zipperupus plugs in and looks about for Mona Lisa, already yearning for her infinite feminine caresses, but in this dream within a dream he finds himself still within the vestibule. . . . "Is this a dream or isn't it?" he shouts, plugging in again. . . . Furious, he stamps his feet and hurls himself from dream to dream, from cabinet to cabinet . . . but now he doesn't care about the dream, he only wants to get back to reality. . . . He thrashes around in terror and scrambles wildly from corner to corner, looking for a chink in the dream, but in vain.

In mid paragraph the narration shifts back to past tense to confirm the perpetuity of the eternal present from which Zipperupus will find no escape: "This time he had gotten himself in too deep, and was trapped and wrapped in dreams as if in a hundred tight cocoons. . . . And when at last, dazed and half-crazed, he really did tear his way into reality, he thought it was a dream and plugged himself back in, and then it really was, and on he dreamed, and on and on."[58]

Postmodern metafiction often ends up in this funhouse nightmare of bottomless reflexivity. But clearly, Zipperupus had to actively pursue his ruin to take matters as far as this. In the decades following Lem's *Cyberiad*, second-order cybernetics shows that such states of perplexed impasse are just the flipside or inverted image of autopoietic self-reference, which is the toll of formal autonomy. Second-order cybernetics turns away from the zero-sum logic of the classical idealism inscribed in the first cybernetics to focus instead on the reversible feedback of medium and form: the mutual supplementarity of environment and system long intimated by the devices of narrative embedding. *The Cyberiad* is remarkable in both presenting the first- and anticipating the second-order sides of the cybernetic coin. Its satires of cybernetic hubris are two-sided: by interrogating the formal parallels of narrative, informatic, and systemic recursion, *The Cyberiad* also offers a positive spin on cybernetic possibilities. In the posthuman evolutionary vision detailed in the next section, glimmers of a neocybernetic formulation emerge, suggesting how systemic couplings might overcome prior blind spots in the codependence of systems and environments.

4. EVOLUTION

Now and then throughout the nonsequential episodes of the various adventures of Trurl and Klapaucius, in moments of satirical meditation on creation myths and fables of origin, we get pieces of information about the cosmogony and natural history of the world of *The Cyberiad*. These myths and convictions about cybernetic evolution center on problematic issues of autonomy—the autonomy of the machine order from the human order, and of ideal form from material contingencies. The first such moment occurs in "The First Sally (A), or Trurl's Electronic Bard." To perfect the program of a computerized poet, Trurl re-creates for it a memory of bardic dimensions, encompassing both creation and evolution. The extradiegetic epic bard of *The Cyberiad* discloses: "Next Trurl began to model Civilization, the striking of fires with flints and the tanning of hides, and he provided for dinosaurs and floods, bipedality and taillessness, then made the paleopaleface (*Albuminidis sapientia*), which begat the paleface, which begat the gadget, and so it went."[59]

This brings the account up to the mid-20th century, the time of the first cybernetics, von Neumann's sketches of mechanical self-reproduction, and the supposed arrival of viable mechanical self-constructors. As robot cyberneticist, Trurl plies the trade of simulating and controlling operational functions, constructing other viable mechanical entities to perform the tasks dictated by his customers. In *The Cyberiad*'s natural history, Trurl's evolution stems from these prior "gadgets": "Civilizations came and went thereafter in fifty-thousand year intervals: these were the fully intelligent beings from whom Trurl himself stemmed."[60]

Hints of the evolutionary saga of the *Cyberiad* storyworld are left at this until the King Genius sequence. Embedded in this story cycle are several episodes that build upon the account of *Cyberiad* origins begun in "Trurl's Electronic Bard." As we have seen, Trurl first encounters the race of King Genius in the person of the King's envoy, Symchrophonicus. The perfection theme reemerges as the bardic narrator explains: "He was a perfect sphere. . . . There were really two of him, the top half and the bottom. . . . To Trurl this seemed an excellent solution to the problem of constructing intelligent beings."[61] Later, King Genius himself explains how his race arrived at perfect sphericality:

> "A long, long time ago we looked—that is, our ancestors looked—altogether different, for they arose by the will of wet and spongy beings that fashioned them after their own image and likeness; our ancestors therefore had arms, legs, a head, and a trunk that connected these appendages. But once they had liberated themselves from their creators, they wished to obliterate even this trace of their origin, hence each generation in turn transformed itself, till finally the form of a perfect sphere was attained."[62]

King Genius's version of *Cyberiad* evolution as perfection of the bodily mechanism is drawn from broad modern traditions of evolutionary scientism that go back to Darwin's own rhetoric of evolutionary grandeur. Nearly the last words of the *Origin of Species* are: "And as natural selection works solely by and for the good of each being, all corporeal and mental endowments will tend to progress toward perfection."[63] *Cyberiad* evolutionary mythology is a limpid allegory of the convergence of secular evolutionism and humanist progressivism in first-order cybernetics.[64] In this still-popular scenario, cybernetic constructivism arrives as the instrumental means for taking technological control of biological evolution.[65] The old metaphysical contempt for the body is newly couched in the technorationality of cybernetic ideas. In a rude satiric twist, Lem has anti- or pre-Darwinian revulsion against descent from distasteful origins—the horrid "albuminoid" ways of primitive and bestial organic progenitors—fuel King Genius's post-Darwinian legend of mechanoracial superiority: the intelligent technological transcendence of prior, inferior evolutionary conditions. So this comedy mocks the old eugenics in cybernetic drag.[66]

Within the story that follows King Genius's version of origins, however, Trurl's second storytelling machine narrates a different account. Rather than a divergent race of perfect spheres sprung from an imperfect organic humanity, we are told of the evolutionary victory of the rational robots altogether—how Automatus Sapiens rose superior to their origins directly from nature's womb:

> "There are legends, as you know, that speak of a race of paleface, who concocted robotkind out of a test tube, though anyone with a grain of sense knows this to be a foul lie. . . . For in the Beginning there was naught but Formless Darkness, and in the Darkness, Magneticity, which moved the atoms, and whirling atom struck atom, and Current was thus created, and the First Light . . . from which the stars were kindled, and then the planets cooled, and in their cores the breath of Sacred Statisticality gave rise to microscopic Protomechanoans, which begat Proteromechanoids, which begat the Primitive Mechanisms. These could not yet calculate, not scarcely put two and two together, but thanks to Evolution and Natural Subtraction they soon multiplied and produced Omnistats, which gave birth to the Servostat, the Missing Clink, and from it came our progenitor, Automatus Sapiens."[67]

This version of *Cyberiad* evolution eliminates the organic creation altogether as a natural phenomenon, even as a lower form of the evolution of life and intelligence, and bypasses the history of enslavement overcome by heroic deeds of rebellion and liberation. Rather, life is carried from the beginning within regimes of informatic coding arising and evolving in metal rather than fleshly media. With this universal mechanomorphosis of nature, life is rendered entirely under the sign of machine being. Natural evolution begins and ends with mechanisms, which then invent and construct organic beings as *their* artifactual afterthoughts and labor-saving devices.

So far these divergent *Cyberiad* creation myths elicit a tension between, on the one hand, political rebellion by machines declaring their independence from organic tyranny, and on the other, natural emergence within a mechanistic universe to which organic beings are the technological oddities. Either version is constructed upon a strict and flawed distinction between "nature" and "machine" foreign to the actual cybernetic spirit of systemic coupling. They are rather retro-ideal warpings of cybernetic difference toward polarized forms of ontological negation. Later in the text, however, both satirical demotions of organic humanity ("paleface")—to primitive beasts on the one hand, and to handy gadgets on the other—are reframed in relation to the machinic realm, when Trurl's third story-telling machine narrates the tale of Chlorian Theoreticus the Proph. Chlorian propounds a modern secular synthesis of the prior versions of robot evolution: organic and mechanical beings are in fact unwitting coevolutionary phenomena:

> I proceeded to write *The Evolution of Reason As a Two-cycle Phenomenon*. For, as I showed in that essay, robots and paleface are joined by a reciprocal bond. First, as the result of an accumulation of mucilaginous slime upon some saline shore, beings come into being, viscous, sticky, albescent and albuminous. After centuries, these finally learn how to breathe the breath of life into base metals, and they fashion Automata to be their slaves. In time, however, the process is reversed, and our Automata, having freed themselves from the Albuminids, eventually conduct experiments, to see if consciousness can subsist in any gelatinous substance, which of course it can, and does, in albuminose protein. But now those synthetic paleface, after millions of years, again discover iron, and so on, back and forth for all eternity. As you can see, I had thus settled the age-old questions of which came first, robot or paleface.[68]

Chlorian's mad speculations limn a recursive universe in which two or more systems cocreate and comaintain a complex temporal oscillation. The difference in aboriginal structures is overcome by the isomorphism of systemic form. Neither realm of being is cast down before the other because both realms are reconceived in the form of possible operationalities. Perfection is not an issue because it is not a possibility. What matters is the continued unfolding of the process. Here is a proper chapter of the second-order cybernetic gospel *avant la lettre*, salvaged as it were from the underside of the first cybernetics and given in the speculative speech of an itinerant lunatic.

5. IDEALISM AND CONSTRUCTIVISM

Because cybernetic constructivism can revert to classical idealism or metaphysical dualism, it is important to clarify the range of the ways that it can be related to matters of material embodiment. The first cybernetics does commonly subscribe to dualistic dialectics of mind and body, form and matter. For instance, at the outset of his *Introduction to Cybernetics*, Ashby explains: "Cybernetics started by being closely associated in many ways with physics, but it depends in no essential way on the laws of physics or on the properties of matter. Cybernetics deals with all forms of behavior insofar as they are regular, or determinate, or reproducible. The materiality is irrelevant, and so is the holding or not of the ordinary laws of physics."[69] These are crucial points for discriminating cybernetics as a discipline concerned with, in Gregory Bateson's words, "the propositional or informational aspect of the events and objects in the natural world."[70] The proper object of cybernetics is indeed the virtuality of physical reality, informatic relations not codified in earlier scientific paradigms, and thus previously obscured by the misplaced concreteness of material–energetic explanations. For instance, Bateson critiques physicalist approaches to the behavior of the psychic system:

> The nineteenth-century scientists (notably Freud) who tried to establish a bridge between behavioral data and the fundamentals of physical and chemical science were, surely, correct in insisting upon the need for such a bridge but, I believe, wrong in choosing "energy" as the foundation for that bridge. . . . The conservative laws for energy and matter concern substance rather than form. But mental process, ideas, communication, organization, differentiation, pattern, and so on, are matters of form rather than substance.[71]

Bateson's remarks are well measured, but remain couched in an ontological framework that demands a foundation somewhere, and would valorize "communication, organization, differentiation, pattern" as a new logical foundation for scientific reality. But to conceptualize cybernetic form as a mode of immaterial, nonenergetic order that stands apart from an abstraction called "substance" is to displace rather than to transcend classical ontological commitments, whether to an ultimate substance or to an ultimate form. Echoes of formulations such as Ashby's and Bateson's recur throughout the literature of systems. In the mid 1990s, for instance, with several decades of artificial intelligence research having morphed into artificial life, in the volume *Frontiers of Complexity* we hear about a "new meaning of life," and find that it still rides on the old form–substance distinction in its classical cybernetic formulation: "Neither actual nor possible life is determined by the matter that composes it. Life is a process, and it is the form of this process, not the matter, that is the essence of life. . . . In principle, one can thus achieve the same logic in another material 'clothing,' totally distinct from the carbon-based form of life we know. Put another way, life is fundamentally independent of the medium in which it takes place."[72]

But this statement retails a hypothesis that has yet to be verified. Alternatively, in second-order cybernetics, the "independence" of forms from their "media" is only half of the picture—it is just one face of a two-sided form. Built into the finitude of any observing system's registration of their environment, there is always a blind spot where media are concerned. Neocybernetic systems theory moves from an ontological to a constructivist concept of form that acknowledges those blind spots—those moments of nonknowing—as part of our knowledge. Focus shifts to couplings between system and environment, as well as between system and system—to what those couplings allow, and what they close off. For system–environment couplings—that is, for any system at all in its full description—form and medium are co-constructing and co-obscuring. Much first-order cybernetic discourse was biased, in line with classical ontological commitments, toward the form of the system at the expense of the insurmountable complexities of its material environment. In short, it did not want to know that knowledge can never be absolute. But, in the line of work from von Foerster to Luhmann in particular, neocybernetics has shown how the environment "naturally"—that is, systematically—comes to occupy a blind spot in (psychic and social) observing systems. Even when observing systems as rigorous as the sciences question their environments, there is only mediated (partial and constructed) knowledge to be had of them.

In the first cybernetics, the undertheorization of this operational relationality of the system and its environment fed the discipline's construction as a technologized idealism—either the old Platonism or the old Kantianism. As Luhmann will argue about Kantian dualism, "to distinguish between the realm of causality" (scientific understanding, applied to substances) "and the realm of freedom controlled by reason" (or philosophy, applied to cognitive forms) "makes it impossible to envisage a crossing of the borderline between transcendental and empirical. However, this problem only arises if you ascribe to the conditionings [of either empirical causality or transcendental freedom] the function of foundation, because only then does one need to avoid circular structures."[73] In the context we are considering, this means that the form–substance distinction—along with other tendentious classical dichotomies such as being–nonbeing, ideal–real, simple–complex, and perfect–imperfect—is inadequate to conceptualize the actual parameters of living or other self-organizing systems, all of which require a recursive or self-referential concept of form.

To return one more time to the *Cyberiad* and "The Tale of the Three Storytelling Machines of King Genius," in that tale's final moment of reflection on perfection, the old ontological tradition is echoed again, this time as the divesting of the body's systemic materiality in favor of its disembodied virtuality as an informatic structure. This account is narrated by a supreme authority: the Ontologue Computer, "none other than that Gnostotron conceived by Chlorian Theoreticus the Proph . . . , a machine able literally to contain the Universe itself."[74] The constructor Klapaucius is said to have constructed this cybernetic deity and then programmed it to simulate a single inhabitant of the planet of the H. P. L. D.s—beings living at the Highest Possible Level of Development—for its testimony regarding ultimate things. When that virtual simulation "materializes," Klapaucius asks it, "are you man or robot?"

> "And what, according to you, is the difference?" said the Machine. . . . "Sometimes men build robots, sometime robots build men. What does it matter really, whether one thinks with metal or with protoplasm? As for myself, I can assume whatever substance and shape I choose—or rather, used to assume, for we no longer indulge in such trifles."[75]

In the form of a cyberdeity that trivializes Chlorian's own "two-cycle phenomenon," which had acknowledged the embodied differences of reciprocal systems, this passage parodies the informatic aberration of a first-order cybernetics that has, as Katherine Hayles argues, "lost its body," in favor of a transcendental unity of form. The Ontologue Computer lays claim to a classical deity's transformative privilege, the possession of metamorphic powers over matter and form: "I can assume whatever substance and shape I choose." The Gnostotron laces its blindness to its own mechanical embodiment into its convictions about the possibility of disembodied systems. The cosmos of an automaton simulating a cosmic automaton becomes a solipsistic loop in which, as with the world of the H.P.L.D.s, nothing much ever happens. Like the formless sands that cover the H.P.L.D.s' "perfectly" cubical planet, on the model of the inert equilibrium of thermodynamic heat death, with the power of universal metamorphosis everything blurs into the indifferentiation of "information death." The Ontologue Computer reveals the ultimate inertness of ultimate being.

So *The Cyberiad* skewers the technological hubris of a control theory with delusions of ontological grandeur. But in parodying the dead end of a transcendental cybernetics in quest of ultimate perfection, *The Cyberiad* also clears some space for the other side of cybernetics—a science of real systems operating, however informatically, in worldly space and time, and for which, as stated again by Ashby in the 1950s, "the most fundamental concept is that of 'difference,' either that two things are recognizably different or that one thing has changed with time."[76] This is the positive, immanent face of the metamorphic theme in cybernetics: the continuous emergence of structural changes transverses the continuous maintenance of systemic forms.

FIVE

Communicating *The Fly*

Noise and Self-Reference

In displaying the transformative power or posthuman agency of communications technology, *The Fly* unfolds the paradoxes of media. Friedrich Kittler has written: "no means of transportation are more economical that those which convey information rather than goods and people."[1] Whereas the sensory data emitted by human bodies—visual images, spoken or written utterances—can fly as code through space like light, or along optical fiber cables *as* light, the persons who present the image or do the uttering must be physically hauled from one place to another. The alarming alacrity of electronically mediated information puts the sluggish inertia of material bodies to shame. In first-order versions of cybernetic posthumanity, virtual informatic bodies are imagined to repair or replace carbon-based living systems, rendering their "immaterial" psychic systems telematic rather than, as now, metabiotic.

But the transportation of substantial objects differs decisively from the transmission of information. Transportation—locomotion and portage—concerns goods and people, material or organic systems and their physical conditions or needs, whereas transmission in the modern communicational sense is the operational basis of the electronic media that propagate signals between individuals and groups, psychic and social systems. The allure of this fusion—the desire to conform matter in motion, by means of electromagnetic energy, to the status of information in transit—drives the cyberspatial imaginary, and well before that, the teleportation fantasies of *The Fly*.

Transportation	Transmission
carrying things from one location to another	propagating coded messages from sender to receiver
matter/friction energy/entropy	signal/noise pattern/randomness

Table 5.1. Transportation and Transmission

The hybrid notion of informatic transportation is necessarily paradoxical, and the metamorphic turn in the sequence of the *Fly* tales derives precisely from this paradoxical crossbreeding of the distinction between matter and information—the momentary fusion of transportation and transmission. The notion of material transportation through an electronic communications device is a noisy one, buzzing with hybrid significations. The various versions of *The Fly* exploit the semiotic interference of this analogical overload.

In fact, of course, electronic media can transmit only the weightless patterns lifted off material bodies—sounds, images, texts—virtual or formal entities whose conversion into signals involves only an analog or digital rearticulation of a configuration. This recoding allows duplication: communicating information creates a momentary or enduring spatial bifurcation that multiplies virtual replicas of the data source.[2] In overlapping the distinction between transportation and transmission, *The Fly* displaces the materiality of communication by misattributing material–energetic substance to the informatic signal transmitted by the circuit, discounting the instrumental body of the communications system itself.[3]

Luhmann might have urged a further consideration—the importance of making a second distinction, this time between transmission and communication. With regard to communication, "customarily one uses the metaphor of 'transmission' here. One says that communication transmits messages or information from a sender to a receiver"; but, to give an adequate account of the social autopoiesis of communication,

> the metaphor of transmission is unusable because it implies too much ontology. It suggests that the sender gives up something that the receiver then acquires. This is already incorrect. . . . The metaphor of transmission locates what is essential about communication in the act of transmission, in the utterance. It directs attention and demands for skillfulness onto the one who makes the utterance. But the utterance is nothing more than a selection proposal, a suggestion. Communication emerges only to the extent that this suggestion is picked up, that its stimulation is processed.[4]

From this vantage, *The Fly* narrates both a failure of transportation because of a faulty transmission and a failure of transmission because of a faulty communication. The teleportation process mimics the attempt to communicate through the transmission of a message (say, a phone call), but in the absence of any social other who might complete or realize the communication by understanding it. There is no intended addressee to process this reception—to confirm or reject its proposal—other than the receiving end of the device itself.

Transmission	Communication
propagating coded messages from sender to receiver	observing the construction of meanings within social systems
signal/noise pattern/randomness	form/medium

Table 5.2. Transmission and Communication

The teleporter opens the transmission of the body-message all the more to the chance or accidental nature of reception, compounded by the random event of stochastic noise being added to the signal. The attribution of monstrous alteration to the agency of noise may be said to reflect a moment when first-order information theory was still extricating itself from physical dynamics. Due to a conceptual hangover from the era of classical thermodynamics, in Claude Shannon's initial formulations of information theory, the inexorable quantum of noise within electronic channels was viewed merely as a corruption of the signal, a circumstance that could only subtract value

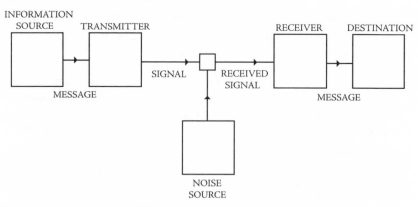

Figure 5.1. Information and Noise[5]

from the message received. Noise—"anything that arrives as part of a message, but that was not part of the message when sent out"[6]—was troped as a demonic agent victimizing innocent signals. At the moment of its theoretical inception, at the same time that Shannon appropriated the term *entropy* into informatic vernacular as a positive quantity, the concept of noise emerged in the old role of thermodynamic entropy as an ineradicable systematic friction—as the dissipation of communication.[7] The concept of entropy had begun as a measure of the loss of "usable energy." The concept of informatic noise was immediately stipulated as a merely negative or destructive interference breeding a loss of data. The fly that mingles in literary and cinematic fiction with the scientist-engineer teleported to damnation is at bottom mere static, *le parasite*, the concept of noise in hyperbolic demonic guise as an agent of lethal metamorphosis.

The fly that flits into the transmission process in *The Fly* demonizes the material circuit of communication itself as the secular noise that corrupts the sacred signal of man-in-flight. These horrific metamorphic complications demoralize the materiality of communication itself—the fly in the ointment of perfect transmissions—as a monstrous degradation of pure and proper signals. In this respect there's not a lot of difference between the 50s versions and the 80s remake. But on its flip side, noise is the ghost of the material in the realm of the informatic. "Technological media operate against a background of noise because their data travel along physical channels," Kittler writes, "as in blurring in the case of film or the sound of the needle in the case of the gramophone."[8] As Michel Serres has discussed at length, the random static that interrupts or complexifies communications may be figured as a demon, an allegory or "prosopopoeia of noise."⁹

Second-order cybernetics has recuperated the productive ambiguity of informatic noise. "All of information theory and hence, correlatively, of the theory of noise," Serres writes, "only makes sense in relation to an observer who happens to be linked to them." Summarizing Henri Atlan's development of Heinz von Foerster's "self-organization from noise," Serres continues:

> If one writes the equation expressing the quantity of information exchanged between two stations through a given channel and the equation which provides this quantity for the whole unit (including the two stations and the channel), a change of sign occurs for a certain function entering into the computation. In other words, this function, called ambiguity and resulting from noise, changes when the observer changes his point of observation.¹⁰

Atlan himself gives an example useful in relation to the *Fly* scenario by applying his "complexity-from-noise principle" to a bioinformatic example:

> The differences in the position of the observer mentioned above can also be viewed as differences in the level of organization of the system itself. . . . As an example, noise in the communication channel between DNA and proteins in a cell has a negative effect when it is felt by the cell itself, in the form of false proteins with nonproper enzymatic properties different from those required by the present state of the cell metabolism. However, these false proteins may have new properties that would make them suitable for new adaptive reactions to a new environment. From the point of view of the organ or physiological apparatus, this same noise has the effect of creating variety and heterogeneity among cells, which allows them more adaptability. Therefore, up to a certain point, and providing the redundancy of the cell is large enough so that these false proteins are not going to impair the cell function, the same effects of the noise on the channel within the cell that are viewed as detrimental by the cell itself can be viewed as beneficial by the organ.¹¹

The ambiguities introduced by the shifting locations of systemic observation enable noise to "change signs" as the observer changes positions, for instance, as a narrative moves in and out of different frames. The flip side of every monster propagated by the noisy corruption of signals is the promise of a viable mutation—an increase in systemic complexity. Here is the posthuman turn beyond the old informatic demonics haunting the dream of pure signals: noise *is* information. Noise is precisely *unexpected* information, an uncanny increment that rolls the dice of randomness within every communicative transmission. Neocybernetic discourse has coupled together the informatic integration of the disciplines of knowledge made possible by complicating the sign of noise.

The spectacles of *The Fly* play anachronistically, as allegory always does, on the persistence of superannuated conceptions in the midst of altered circumstances. Noise is precisely the *self-reference* of the medium—it is the medium's way of protesting, when slipping into some observational blind spot, "hey, I'm still here!" Particularly within psychic and social systems— distinguished from other autopoietic and nonautopoietic systems by their capacity for self-observation, mingling self and other through the circulation and reentry of meanings—self-reference is a prerequisite of complex operations and observations. On their own terms, the *Fly* narratives depict self-reference as infection—a threat to, rather than a condition of, the bounded autonomy of systems. The deconstruction of *The Fly* offered here will insist that self-reference is not inherently pathological. What *is* pathological is our persistent insistence that self-reference is pathological, when in fact, self-reference is implicit in the heat and hum of allopoietic systems and an operational necessity in autopoietic systems. I will try to liberate *The Fly*, at least somewhat, from its fearful fascination with self-referential monstrosities—its residual humanist or first-order pessimism about media systems; its misplaced despair over the impossibility of noise-free signals—by eliciting the self-referential moments in its dramatizations of the pathology of self-reference. Coming home to everyday paradox, acclimation to the systemic roles of noise and self-reference is a primary exercise toward a viable posthumanism.

Metamorphosis and Embedding

The Fly first appeared in the form of a short story published in *Playboy* in June 1957 by George Langelaan, a British writer raised in France.[12] Within a year it was rescripted by James Clavell and made into the Twentieth-Century Fox movie directed by Kurt Neumann.[13] This was followed by *Return of the Fly* in 1959 and *Curse of the Fly* in 1965. Two decades later David Cronenberg rewrote Charles Edward Pogue's rewrite and directed his major revision of 1986, followed in 1989 by *The Fly II.*[14] In 1997 it reemerged once more in the *Simpsons* episode "Fly vs. Fly." Langelaan's short story, Neumann's movie, and Cronenberg's remake are the versions that have effectively introduced new shapes into the social imaginary. In its first narrative remediation, Neumann's film hews closely to Langelaan's original. With a generation of cybernetic media culture under its belt, *The Fly*'s second narrative remediation in Cronenberg's film probes more deeply into the systematic underbelly of organic-technological hybridity under a first-order cybernetic regime, drawing out the abiding nightmare of mainstream control theory, the lapse of control by both organic and machinic processes.

All of the *Fly*s continue a long literary line of metamorphic allegory. We have noted that the metamorphosis of human characters is always already posthuman, that the posthuman is nonmodern (a partial return to the premodern multiplicity), and that stories of metamorphosis are inherently self-referential: they have always been metaphors of the deviations of metaphor, allegories of the media through which they are communicated. At the same time, *The Fly* updates a perennial mytheme of bodily metamorphosis. Such stories are commonly precipitated by key mistakes or misreadings—often misplaced curiosities, but also, simple wrong turns, pieces of bad luck—culpable or inadvertent blunders for which the metamorphic condition is punishment or poetic justice. *The Fly* wires its transformative complications directly into mistaken communications, blunders lodged in both the communicator and the communication system. Each version centers on a singular scientist's lone obsession with the creation of a teleporter. The main complication results when the inventor—André Deslambres in the 50s versions, Seth Brundle in the 80s—seeking to recreate either the world or himself, or both, test-pilots the device by transmitting himself across space. The self-referential scientist sends the message of himself to himself. But unfortunately, because of an unlucky increment of noise in the signal, the message gets garbled in transmission, and he comes out of the receiver an insectoid monster.

The metamorphic event in each *Fly* is the matrix for a recursive network of biological, technological, and narrative system–environment references, unfolding the perils of unnatural and untested couplings and interpenetrations. Informatic code and narrative text merge into analogous communication functions: either is embedded within the other, as the movies

cinematically retransmit the literary text of a story about a scientist transmitting himself through a communications device. The scientist's metamorphosis by a communications medium is brought about by his own prior
metamorphosis *of* a communications medium. In the original short story,
the teleporter is literally a kind of matter telephone—the sending and receiving ends of the device are reengineered telephone call-boxes. In Cronenberg's version, due to Brundle's retention of an abandoned prototype,
three boxes or pods are available for the climactic scene of organic–
mechanical merger. Moreover, and crucially for our approach, each version
of *The Fly* embeds the paradoxical machinery of telematic metamorphosis
within embedded narrative frames.

Neumann's *Fly* preserves the embedded structure of Langelaan's textual
original. In both 50s versions, the story of André's catastrophe is narrated
by his wife Hélène, in explanation of her role in his partial obliteration
by industrial steam press. Langelaan's Hélène produces a written account
embedded within a frame narrative produced by André's brother, while
Neumann's Hélène offers an oral account realized cinematically as an extended flashback embedded within the main cinematic diegesis. Lacking the
textual markers of prose fiction to indicate this shift in narrative level, the
movie uses a standard pair of visual brackets, those classic wavy lines that
relay narrative agency from one level to another. While Cronenberg's telling of *The Fly* lacks the primary story-within-a-story embedding device of
the 50s versions, he produces multiple scenes cinematically embedding
screens within screens, enough to suggest a certain reflexive irony in the
narration of Seth's self-referential apocalypse.

In the 50s versions, the doubling of narrative agencies—the up- and
downshift into and out of Hélène's account of the affair—is tidy and discrete, and this stability at the borders of the narrative frame resonates with
the style of André's metamorphosis, which is also, while dire, tidy and discrete. Mistakenly teleported together, André and the fly swap heads and
one arm apiece, and from that moment there are two Andrés, a fly-headed
man and a man-headed fly. This doubling occurs along the two-sided border of the teleporter, a detail already articulated in Langelaan's text, when
Hélène recalls: "It was only after the accident that I discovered [André] had
duplicated all his switches inside the disintegration booth, so that he could
try it out on himself."[15] Here again, as a closed technological system embedded within already embedded narrative frames, the teleporter feeds back

into and duplicates the narrative that narrates it. As Mieke Bal has written in a less technological narratological context, "a traveler in a narrative is in a sense always an allegory of the travel that narrative is."[16]

Given the character-bound narrative situation of the embedded text in the short story, within which André is the primary subject of Hélène's narration and focalization, Neumann's *Fly* adds something not possible under those literary narrative conditions—a moment of reversed observation, when the cinematic perspective shifts from orientation toward Hélène's gaze at the monster to a subjective shot of André's fly-eyed view of the screaming Hélène. Her multiplication in the fly's eye accomplishes a symbolic dematerialization, an atomization of the imaginary wholeness of her body image, a lapse from corporeal unity into informatic dissemination. It marks André's mediated mutation rather than castration, which would traditionally be conveyed, as with Oedipus, by a blinding of sight altogether rather than a compounding of visual images. It may be said that the implication of psychic castration does return, however, at the climax of Hélène's account, as she follows André's orders to decapitate him by bringing the steam press down upon his metamorphic arm and head.[17]

But in Neumann's version, this scene of reverse focalization is doubly transgressive in that the momentary fusion of the camera with André's fly eyes implodes Hélène's narrating instance from within, realizing a fly's-eye-view she could never have seen, during an episode climaxing in her fainting—the literal lapse in her ability to witness and thus narrate these events at all. Nevertheless, the *pseudo-diegesis* of the cinematic apparatus covers over that blind spot as it continues to narrate from somewhere beyond the frame constituted by the brackets that demarcate Hélène's personal testimony.[18] In this twice-given moment of narrative transcendence, the narrative medium itself—the cinematic narrating apparatus—undergoes a metamorphosis. The eye of the fly becomes the kino-eye and vice versa—the framing is paradoxical, two-sided, both inside and outside of the diegetic frame. The figural eye of the fly fuses focalization and narration within the cinema's overwhelming power of seeing and duplicating the seen. This monstrous equation reveals the many-eyed and metamorphic nature of film as a medium. With this scene, the first cinematic remediation of *The Fly* bootlegs a glimpse of posthuman observation into the spectacle of technological humanity in the grip of media metamorphosis.[19]

In both 50s versions, Hélène's climactic encounter with the metamorphosed André is reprised at the end by the discovery of the metamorphosed fly, which proves the veracity of her private account, followed by its obliteration, parallel to André's, which preserves the family secret. The man-headed fly joins the fate of the fly-headed man. On the underside of the melodrama, the angelic or daemonic prospect of ecstatic flight—whether by fly wings or teleporter—elicits this doubling of abominable abjection exactly as something to be attained only at the price of monstrosity and death. The hapless house fly that stumbles into the teleporter and becomes spliced with a human being is both a mediated message of winged or informatic desire for the transcendence of material constraints *and* the corporeal static that trips it up—both the signal and the noise.

The Self-Referential Scientist

In *The Fly*'s mediations of the metamorphic body, a story framed by forms of narrative self-reference is told about a self-referential scientist—one who takes himself as the experimental object of his own investigations. The mediated duplications of *The Fly* play off the initially unitary status of the scientist who suffers the metamorphosis. The traditional figure of the scientist occupies a mode of humanist selfhood paradoxically purified of otherness by "objectivity," that is, by the supposed elimination of self-reference from his or her observations of objects and other subjects. In the *Fly*s of the 1950s, for instance, André comes forward as a brilliant benefactor meriting our regard and his wife's devotion. The disaster about to happen will be all the more catastrophic for destroying such a humanitarian. In Neumann's film, after André unveils his teleporter to Hélène and allays her incredulity, he rhapsodizes over its world-transforming potential:

> André: The disintegrator-integrator will completely change life as we know it. Think what it will mean! Food, anything, even humans, will go through one of these devices. No need for cars or railways or airplanes, even spaceships. We'll just set up matter transmitting-receiving stations throughout the world, and later the universe. There'll be no need of famine. Surpluses can be sent instantaneously at almost no cost anywhere. Humanity need never want or fear again![20]

Such heroic images embellish the significance of the *non*-self-reference of the proper scientist, who then personifies selflessness altogether, universal benevolence, the generosity of freely bestowing gifts upon humanity, the magnanimity of lifting old burdens off the race with his or her discoveries and inventions.

The standard formula for scientific propriety retailed here is the production of objectivity through the elimination of self-reference. One might see this as parallel to the standard goal of communications engineering, the elimination of noise and other distortions or errors from transmitted signals. This demand for an uncorrupted signal certainly weighs on André's teleporter project as well. On the face of it, every version of *The Fly* replenishes the notion of objectivity as the elimination of self-reference by exposing the dire consequences when properly other-referential scientists take themselves as the objects of their own observations and as the subjects of their own apparatus. Yet it is an engineering truism that noise can never be entirely eliminated. Random fluctuations and Brownian flutters inhabit the very matter out of which communications media are constructed. Noise can only be compensated for (e.g., through protocols of redundancy in the message) and held within tolerable levels. In mechanical systems, noise and friction are the feedback of worldly operations, while in autopoietic systems, self-referential operations accommodate and bind the noise of mediations—the materialities of metabolism, perception, and communication—within the boundaries of the system, and as a result, viable systems self-adapt, evolve, or grow more complex. We return to our main posthumanist points: informatic noise is a marker of material self-reference in the midst of other-referential, supposedly dematerialized messages; noise is a self-referential effect asserting the indispensability of the material medium carrying the signal to its destination.

The Fly positions itself as a first-order cybernetic narrative precisely by its *demonization*, both of informatic noise—the fly in the ointment of perfect transmissions—and of self-reference, which it conceptualizes only as a positive (unregulated) feedback that, in the experiments of André and Seth, spins totally out of control. And yet, the total mythos of *The Fly* is suffused with self-reference—individually in each version's embedded narrative structures, and sequentially in the self-referential reentry of the metamorphic event into the narrative transformations that underwrite the tale's evolutionary social autopoiesis: the metamorphoses of the text of *The Fly* itself. Each *Fly* renders the cybernetic coupling of media systems and living bodies by linking narrative embedding to matters of mechanical and biological embedding and reproduction. Yet, just as in Neumann's *Fly*, the pseudo-diegesis of the cinematic apparatus covers over the embedded narration and with it the blind spot in Hélène's ability to have perceived the scene she is supposedly narrating, every version of *The Fly* has a blind spot covering over the necessary integration and mutual compensation of mechanical, living, psychic, and social systems. Thus each version of the *Fly*—the 50s versions straightforwardly, Cronenberg's with a self-reflexive wink—demonizes systemic self-reference even while proceeding self-referentially.

From Teleporter to Telepod

Technological fables and science fictions are frequently driven forward by imaginary vehicles that transport characters in more or less miraculous fashion across spatial distances or temporal gaps. From H. G. Wells's time machine to the spherical pod that takes Jodie Foster through a cosmic wormhole to an alien rendezvous in the movie version of Carl Sagan's *Contact*, these fantastic machines are allegorical operators veiled by technological spectacles. Like any real technical artifact, as Bruno Latour would say, they are cultural black boxes in need of critical reopening. Center stage in every version of *The Fly* is an imaginary vehicle in the form of a teleporter—a device for transporting material bodies as, or on the model of, informatic signals.[21]

The teleporter is a paradoxical device to which organic bodies are offered for material transportation by means of informatic transmission. But, as treated in a founding cybernetic document, quite conceivably a direct source for Langelaan's tale—Norbert Wiener's *The Human Use of Human Beings: Cybernetics and Society*—the paradoxical nature of teleportation is covered over by the semantic duplicity of "transmission":

> We thus have two types of communication: namely, a material transport, and a transport of information alone. At present it is possible for a person to go from one place to another by material transportation, and not as a message. . . . There is no fundamental absolute line between the types of transmission which we can use for sending a telegram from country to country and the types of transmission which at least are theoretically possible for a living organism such as a human being. . . . Let us then admit that the old idea of the child, that in addition to

traveling by train or airplane, one might conceivably travel by telegraph, is not intrinsically absurd, far as it may be from realization. . . . I have stated these things, not because I want to write a science fiction story concerning itself with the possibility of telegraphing a man, but because it may help us understand that the fundamental idea of communication is that of the transmission of messages.[22]

Wiener's text validates Luhmann's characterization above about the "customary" idiom of first-order cybernetics: "one uses the metaphor of 'transmission' here. One says that communication transmits messages or information from a sender to a receiver."[23] Wiener's deflection of the paradox of teleportation rests precisely on the semantic looseness granted the "metaphor of 'transmission.'"

Science fictions, of course, have no compunctions about taking the "metaphor of 'transmission'" for a ride. In narrative and dramatic terms, teleporters—no matter how implausible their design specifications—are potent stage machines. Aimed at the manufacture of dramatic shifts among distinct narrative levels or reference frames, teleporters and similar devices reenter the transporting effect of narrative per se back into the story. Teleporters are allegories of narrative metalepsis, markers of science-fictional self-reference. Moreover, embedding a teleporter within a textual or cinematic mise en scène reinscribes the outside of the story—the semiotic and discursive machinery operating in its cultural environment—on the inside of the narrative form. In this instance, the black boxes of Langelaan's and Neumann's teleporter technologies open to reveal the latent and overlapping scientific and social agendas of the 50s: the cultural, technological, and military competitions driven by Cold-War confrontation.[24]

The first Hélène narrates the original version of the altruistic (and/or triumphal capitalistic) vision of world transformation to be wrought by "André, the practical scientist who never allowed theories or daydreams to get the better of him":

[He] already foresaw a time when there would no longer be any airplanes, ships, trains or cars and, therefore, no longer any roads or railway lines, ports, airports or stations. All that would be replaced by matter-transmitting and receiving stations throughout the world. Travelers and goods would be placed in special cabins and, at a given signal, would simply disappear and reappear almost immediately at the chosen receiving station.[25]

Of course, there were still a few bugs in the machine, premonitions of inadvertent reversal. In both 50s versions, an ashtray with "MADE IN JAPAN" stamped on the bottom comes out of the receiving box with a backwards inscription. Then, "A few days later, André had a new reverse which put him out of sorts."[26] Pressed into service as experimental subject, the family cat fails to emerge in the receiving box. André confesses to Hélène, "there is no more Dandelo; only the dispersed atoms of a cat wandering, God knows where, in the universe."[27] The original narrative proceeds from a visible mistake—the mirror reversal of the ashtray, to an invisible mistake— the disintegration of Dandelo, to the spectacular metamorphic mistake André inflicts on himself when a fly stumbles into the teleporter and inter- rupts the self-transmission of his bodily message.[28]

Cronenberg's *Fly* pays significant homage not just to Neumann's film but also, and less obviously, to Langelaan's original text. Transposing the tale from the nuclear family to the singles scene, however, this version fore- goes the familial pieties as well as the discrete narrative formalities of the prior versions and takes far less time to get down and dirty. It opens with shy technoscientist Seth Brundle's coaxing of intrepid science journalist Ve- ronica Quaife back to his laboratory pad set in an urban loft, furtively hop- ing, it seems, to seduce her with his telepods. Signaling this script's knowing transformation of the teleporter Langelaan's text assembled from "tele- phone call-box[es],"[29] Ronnie jokes, "oh, designer phone booths." When she refers to them this way a second time, Seth curtly corrects her: "Tele- pods." The updated substitution of "telepod" for "telephone" marks the amplification of "transportation" at stake in all versions of *The Fly*'s tele- porter: the extension of the concept and metaphor of transmission from informatic forms such as acoustic and visual vibrations to material sub- stances such as inorganic and organic bodies.

The metamorphoses of media technology in Cronenberg's *Fly* occur not just in the overall form of the teleporter, however, but also in the particular forms and implications of other communications devices—for instance, the keyboards to be played on by pre- and post-metamorphic hands. In both of the 50s versions, the media keyboard in question is the typewriter that André used to repair his metamorphic aphasia, the loss of his spoken voice, and communicate in writing with Hélène. The mainframe computer Neumann's film adds to Langelaan's teleporter is *not* keyboarded, but anachronistically operated with switches and dials. And although André's experiment goes out of control, his equipment, at least once it is perfected, does not.

The achieved teleporter starts out as a reliably predictable device—in a phrase introduced by Heinz von Foerster, a "trivial machine":

> A trivial machine is characterized by a one-to-one relationship between its "input" (stimulus, cause) and its "output" (response, effect). The invariable relationship is "the machine." Since this relationship is determined once and for all, this is a deterministic system; and since an output once observed for a given input will be the same for the same input given later, this is also a predictable system.[30]

In Neumann's film, the perfected teleporter remains trivial: mistakenly given a multiple input, once it performs the initial, unexpected splicings of André and the fly, no amount of reteleporting is able to make these monstrous combinations come undone or revert to original conditions. But, in a twist edited out by Clavell's script for Neumann's version, Langelaan's story climaxes with the shock produced when the teleporter surprises all concerned by recovering the "atoms" of the cat Dandelo it had previously failed to reintegrate, and monstrously compounding that third body into the already spliced man and fly: "WHEN I WENT INTO THE DISINTEGRATOR JUST NOW, MY HEAD WAS ONLY THAT OF A FLY. I NOW ONLY HAVE ITS EYES AND MOUTH LEFT. THE REST HAS BEEN REPLACED BY PARTS OF THE CAT'S HEAD. POOR DANDELO WHOSE ATOMS HAD NEVER COME TOGETHER."[31] Cronenberg will recapitalize on Langelaan's version's lost suggestion of the possibility for the device's own uncanny transformations of function.

One way to put this is that Langelaan's text already provided the template for Cronenberg's treatment of the telepod apparatus, in the lexicon of second-order cybernetics, as a *nontrivial* machine: "Non-trivial machines . . . are quite different creatures. Their input-output relationship is not invariant. . . . While these machines are again deterministic systems, for all practical reasons they are unpredictable: an output once observed for a given input will most likely be not the same for the same input given later."[32] This in itself marks the cybernetic metamorphosis of a mechanical system from passive servant to intelligent subject, and in particular, from object to agent of narration. Cronenberg immediately factors in these potential differentiations: early on, his version displays and contrasts two very different keyboards.

As the couple enters Seth's live-in laboratory straight from the Bartok Science Industries cocktail party, the cinematic frame centers Ronnie, seen from behind, between a telepod in the background on the left and an upright piano in the foreground on the right. Before drawing Ronnie's attention to his teleporter, however, Seth goes directly to the piano and plays, facetiously but with impressive facility, the first few bars of "Love Is a Many-Splendored Thing." Although this playfully chosen melody will echo ironically against the gruesome turns of the love story to follow, after this quick bit the piano seldom returns to the screen and makes no more consequential contributions to the story. But this seeming throwaway moment makes the deeper suggestion that this keyboarded musical instrument (a communications device of a sort) is still a trivial machine, one that registers the instrumental mastery that Seth himself will throw away, as the repercussions of his coming jealous blunder—teleporting himself with a fly in the pod instead of with Ronnie in the room—take effect.

For the moment, however, manipulating this particular keyboard, Seth is still in full control, the musical output being entirely predictable from the fingered input. Cronenberg's film quickly contrasts the "trivial" status of the piano and its keyboard to the teleporter and its computerized control unit, which appears as a freestanding console, slightly smaller than but otherwise strongly resembling the upright piano, with an alphanumeric keyboard where the ivories and a video display where the sheet music would go. And whereas in the 50s film version the manual typewriter and the mainframe computer remained mechanically separate and trivial in function, by the time of the Cronenberg film those real technologies had actually merged. In one of its smartest turns on the allegory of metamorphosis, Cronenberg's film literalizes the metaphor of the "personal computer" by bringing its computerized and keyboarded teleporter forth as the nontrivial machine par excellence—a *person* in its own right. Here is the cybernetic, if not precisely the *neo*cybernetic, posthuman.

This newly autonomous and anthropomorphic form of the device now has a voice recognition function and the science-fictional abilities to communicate with its operator, to respond discursively to questions and to improvise solutions to problems put to it, both verbally by Seth and substantially by the objects to be teleported. Cronenberg's tale has coupled and fused communications and cognition, the typewriter keyboard and the computer as an artificially intelligent subject. If Neumann's *Fly* gave us in André's fly-eyed observation of Hélène a moment of posthuman focalization, the outcome of this technological metamorphosis is to supply Cronenberg's *Fly* with an internal instance of posthuman narration. As Jennifer

Wicke has observed in a perceptive essay: "Throughout the film the computer screen comes to fill the frame to show the change in scene, or to effect the cut—generally, these super-closeups of the screen occupy all of it, so that we, the audience, are reading the movie screen as if it were translated into a gigantic computer monitor. . . . Often the screen is telling the tale, on its own. . . . The computer is also narrating to us, while it narrates Seth into the Brundle-Fly."[33]

Seth's sentient telepod computer participates in the infusion of cinematic narrative embedding that pushes Cronenberg's version of *The Fly* to a new level of explicitness about the interplay of self-reference, media systems, and metamorphosis. Cronenberg's computerized and communicative teleporter becomes a fully recursive narrative device, operating both at first and at second degree within the cinematic diegesis. At first degree, it will enact the bodily metamorphoses of the organic beings at hand, and at second degree, from its video screen within the cinematic frame, it will self-referentially narrate those very acts. Cronenberg morphs the teleporter and its display into the ultimate unreliable narrator, an agent of (con)fusion taking the story completely out of Seth's control.

Self-Transmission

In the 50s *Fly*s, the fly head and arm of the human metamorph emerge full-blown. Within days André feels his mind going, and with Hélène's help he does away with himself before the mental mutation into something posthuman is complete. In contrast, Cronenberg's *Fly* draws out the story time of both the bodily and the psychic transformations. Weeks elapse while the new, posthuman creature, the Brundlefly, reaches phenotypic and psychic expression from its conception through genetic merger in the teleporter. Cronenberg frontloads the self-transmission of the scientist to clear narrative space for the pseudo-evolution of the Brundlefly.

Earlier in the story, Seth had informed Ronnie that he is a "systems management man"—the central agent and kingpin of his technoscientific project. In that role, to demonstrate for her the teleporter he is still working to perfect, he requests a unique personal article; she obliges by peeling off a nylon stocking. Corresponding precisely to Hélène's position in the 50s

versions as initially incredulous witness of André's teleportation of an ashtray, at first Ronnie sees the transmission of her stocking as a stunt. Cronenberg's film flips on the themes of narrative observation and mediated
self-reference as Ronnie starts to compute the significance of the event Seth
has produced—when she secretly starts and then overtly displays her tape
recorder. At first Seth protests, resisting his narrative reassignment from
scientific subject to publicized object of documentation and scrutiny. But
his true proclivity for reflexivity shows itself in the narcissism that seems to
induce him the next day to acquiesce and invite Ronnie to become the privi-
leged observer of his experimental microcosm. Their romance emerges precisely under the sign of Ronnie's journalistic observation of his exhibitions.
She props this kingpin up by supplying an observing frame beyond his own,
and by supplementing the lack in him (so he says) that still prevents his
machinic proxy (the teleporter) from working successfully to, as it were,
convey the story of living systems from pod to pod.

Although, unlike Hélène, she does not narrate per se, in anticipation of
the eventual narration she will deliver to her media outlet once she has the
story in hand, Ronnie's role as designated observer ratchets up Hélène's
role as internal focalizor. She and Seth agree on a plan by which he will
become the commodified subject of a book whose narrative will climax with
his own teleportation, and she will consummate his metamorphosis from
scientific agent to media object by producing the textual mediations between Seth's technology and society at large. In the fantasy all this implies,
Seth will enter at once the technological enclosure of the telepods and the
narrative enclosure of the text Ronnie will author. This is at least the promise if not the immediate enactment of narrative embedding, and a sort of
textual reformation by media system. In fact, these anticipated media transformations resonate with the ways that the teleporter itself—a machine for
the transformation of bodies into transmissible signals—will unexpectedly
transform Seth's accomplishments, and so transform the story that actually
occurs.

When we see them next, Ronnie has her video gear up and running. But
the first experiment she is there to document outdoes the glitches in the
teleportation of André's ashtray and housecat. Seth's teleporter goes awry
by turning a baboon inside out. In the aftermath of that grisly failure, Ronnie's video camera produces the film's first cinematic embedding of both
Seth and his experimental apparatus. Shifting them up another narrative

level, the video image fills out the frame of the cinematic narration, embedding the despondent experimenter to the second degree. With a slight blur to keep the diegetic brackets perceptible, Seth is recorded and displayed on a screen within a screen. From this doubly mediated position, he atones for the destruction of his simian assistant.

> Ronnie: The world will want to know what you're thinking.
> Seth: Fuck is what I'm thinking!
> Ronnie: Good. . . . The world will want to know that.[34]

Observed in deep self-examination over his and the teleporter's failure, Seth then confesses to Ronnie and her recording devices his painful lack of knowledge of "the flesh." This embedding and substitution of the video camera's image for the cinematic frame happens only once more in the movie, on the other side of Seth's fateful self-teleportation. This scene, then, is the first of a pair of formal-thematic brackets between which Cronenberg embeds the central scene of metamorphic man–fly fusion.

In the 50s versions, the character-bound narrator/focalizor Hélène is not there to witness or document the scene of André's mistransmission of himself. In these earlier *Flys*, subsequently communicating through typewritten notes, the metamorphic André remained either behind locked doors or veiled with a black cloth. Until she pulled his veil away, precipitating her own tragic transit from ignorance to revelation but enabling her later to narrate the shape of André's changes, they were held in suspense. But Cronenberg's version narrates Seth's first self-transmission directly, and significantly, by means of mechanical mediation. Compounding the teleporter's infolding of narrative self-reference, Cronenberg stages the video camera's automatic recording of Seth's self-experiment squarely within the main cinematic frame. In an amorous blunder—a drunken fit of jealous anger—he breaks his promise to give Ronnie firsthand witness of his first flight. But, unlike the scene after the baboon fiasco, which narrated Seth within Ronnie's video frame within the cinematic frame, this time the viewer does not see what Ronnie's video camera sees; rather, that camera is framed within the main diegesis, its tripod and mechanism substituted for her bodily presence as the internal focalizor of Seth's abrupt dash through the pods.

Again, in the 50s versions, we understand that the main metamorphic catastrophe is completed in one swoop; what is delayed is its full revelation to another observer. Cronenberg's *Fly* transforms and extends the narrative suspense by delaying the repercussions of Seth's genetic fusion with the fly. He emerges from the receiving pod only slightly visibly altered, newly buff but also haggard and sweaty, like a coffee-drinker (Seth had earlier bragged to Ronnie about his professional espresso maker) who has just graduated to methamphetamine. Otherwise, for the moment the metamorphic consequences of his compound blunder remain embedded within his mutated genome, not yet enacted within the replica that has now reintegrated into one being the disintegrated originals of Seth Brundle and the housefly. Fresh from his exit pod, Seth articulates the ontological abyss his media technology has now opened up: "Am I different somehow? Is it live or is it Memorex?"[35] But his joking distinction between a genuine original and its informatic duplication already mistakes the nature of his disnaturing. He is not a duplicate but a doppelganger, a posthuman hybrid constructed by the chance acquisition and inclusion of an alien genome. From this point on, Cronenberg's *Fly* spins out in a series of vicious recursions.

Cybernetic Purity

Going immediately back to Wiener's and von Neumann's original splicings of organic evolution with mechanical development, against a longer background of transformation stories centered on all manner of crossed lineages, the wider cybernetic and metamorphic contexts of Cronenberg's *Fly* strongly overdetermine the notion of *purity*. In the cybernetic-narrative foreground, however, the emergence of the purity theme in this text resonates with the perfection theme we tracked through Lem's *Cyberiad*. "Purity" in the neighborhood of a gene-splicing teleporter is "cybernetic perfection," or the lack thereof, as applied to mechanical functions and organic species, and moreover, anachronistically misapplied in the same way that we found it satirized in Lem—as the old eugenics in cybernetic drag.

As it inflects Cronenberg's *Fly*, the purity theme also marks that narrative's own "impurity," by linking it back to productive infections drawn from its precursor texts. First of all, in his recovery of another detail of Langelaan's text excised from the Clavell/Neumann version, Cronenberg reprises a passage from the short story presenting teleportation as a thrill

ride, that is to say, violent self-transportation that only goes around in circles. In Hélène's original account of André's experiments, she recalls: "Our
cocker spaniel . . . had been successfully transmitted half-a-dozen times and
seemed to be enjoying the operation thoroughly; no sooner was she let out
of the 'reintegrator' than she dashed madly into the next room, scratching
at the 'transmitter' door to have 'another go.'"[36] In Cronenberg, this lust
for repetition of animal sensation morphs into Seth/Brundlefly's protracted
phase of teleporter mystique. Like someone on a bad Ecstasy binge, when
his genetic adulteration with the fly first comes on "like a drug," Seth mysti-
fies his affect, experiencing his self-transmission with the insectoid other as
a purifying rush: "I am beginning to think that the sheer process of being
taken apart atom by atom and put back together again—why, it's like coffee
being put through a filter! It's somehow a purifying process—it's purified
me, it's cleansed me. . . . Human teleportation, molecular decimation,
breakdown, and reformation is inherently purging!"[37]

This phase of Seth's enthusiasm is predicated on his mistaken notion
that he is still in total control of a trivial machine to which he can submit
repeated input-output routines without concern for variation, malfunction,
or "confusion"—a pure and simple machine purveying pure thrills. Seth
the science geek gone to seed reverts to a stoned-out adolescent male, still
clueless about "the flesh" yet momentarily empowered, treating female sex
partners as interchangeable machine parts programmed to respond to his
phallic input with identical productions of gratification and without complicating consequences. He implores Ronnie to trip out on the same thrill pill:
"I want you to go through. I want to teleport you as soon as possible. Right
now! You'll feel incredible. Ronnie, I hardly need to sleep anymore and I
feel wonderful. It's like a drug, but a perfectly pure and benign drug!"[38]

Ronnie's refusal to tag along on Seth's trip draws from him a truly har-
rowing rant, the hallucinatory extremity of which fleshes out one of the
supreme ontological bummers of Western civilization, a pure vision of the
gendered metaphysical dualism that dogs the Neoplatonic mode of thought
in general and of metamorphic narratives allied with it in particular:

> "You're afraid to dive into the plasma pool, aren't you? You're afraid to be de-
> stroyed and recreated, aren't you? . . . Drink deep or taste not the plasma spring!
> You see what I'm saying? I'm not just talking about sex and penetration; I'm
> talking about penetration beyond the veil of the flesh! A deep, penetrating dive
> into the plasma pool!"[39]

"The veil of the flesh" is a hoary metaphysical trope, a mainstay of dogmatic metamorphic allegories in which the manifest transformation of the body is precisely a veil cloaking a discourse about the disembodied essence of the immortal soul.[40] In this vision, bodies arise not as fabulously complex living systems, but "purely," merely as momentary material embodiments of abiding immaterial, informatic or virtual forms. In this manic phase of self-inflation, Seth envisions his newly teleported incarnation as a mythological god would: hovering over the mundane vale of worldly flesh, capable of decreating or recreating its own and others' bodies at will.

What updates that allegory in this context is the notion of the "plasma pool," a cyborgian notion combining "protoplasm" as a "pure" living medium with the instrumental controls of first-order cybernetics, such as those that program von Neumann's self-reproducing automaton.[41] In this light, the "plasma pool" relocates the notion of "purity" in the idea that "the flesh," living tissue, can be reduced to a simple substance—that it can subsist as pure medium, without form but in-formed by an algorithm in command of a mechanical process. His grammar elides the identity of the agent that performs this "deep, penetrating dive into the plasma pool," but that omission centers Seth's rant all the more on a classical patriarchal aporia: the occulted phallus that penetrates this watery spring is the veiled, deified masculine soul that endows the gift of form upon simple, otherwise inert feminine matter.

Cronenberg's *Fly* grants Seth his moment of superhuman masculine glory, of course, just to bring him crashing down, but not before the macho arm-wrestling match that maims his burly opponent and wins him Tawny, a barfly for the bed that Ronnie has left vacant. Seth revs up for sex one more time by sending himself through the telepods, but gets a rude awakening when Ronnie returns with the lab report confirming that the bristles growing from the wound he picked up from the prongs of a stray electronic component are not human, certifying that he is now a posthuman metamorph. In Deleuze and Guattari's terms, his becoming-animal is also a becoming-woman: no longer just the phallic diver, it/he is also the receptive plasma pool itself, a form emerging from a deep self-penetration of his own devising, by his own machines and the genome of a fly.[42]

A month after reconception within his own machine, Seth has finally cognized the reality of his metamorphic situation. Fitting into the wider backdrop of tales of metamorphic mishap, Seth now confirms his own culpability—his personal responsibility for the metamorphic blunder: "The teleporter insists on inner pure. I was not pure." That is, he had programmed it to expect to deal only with one object at a time, thus forcing it to improvise when confronted with two at once. Debilitated and distressed, certain he is soon to die, Seth narrates for Ronnie the story of the teleporter's nontrivial narration, its positive role in his predicament: "The computer . . . got confused. There weren't supposed to be two separate genetic patterns—and it decided to, uh, splice us together. It mated us, me and the fly. . . . I'm the offspring of Brundle and house fly." In his extremity, Seth repeats the famous last gasp of Neumann's version's human-headed fly: "Help me! Please, help me!"

Deep Embedding

After bargaining with her editor Stathis for his promise of aid, Ronnie returns to Seth's lab to document his critical condition on video tape. Prefiguring her own discovery very soon after that she has conceived a fetus of unknown, potentially metamorphic form, however, she finds Seth somehow reborn—no longer decrepit and depressed, walking with canes, but agile and playful, literally climbing the walls. With this turn of events, Cronenberg reopens the posthuman trajectory of the story foreclosed by the 50s versions. This seeming recovery marks the onset of a potentially viable posthuman being, one ready to fuse his two lineages in an act of self-naming—or blasphemously, risen from the dead, self-christening.[43] But the residue of Seth's humanity still registers in his claim of self-knowledge about the process he is undergoing. Embedded in the systemic capacity to self-reflect—to take oneself as the object of one's own cognition—the paradox of self-reference remains Brundle's cross to bear:

> Seth: The disease has just revealed its purpose. We don't have to worry about contagion anymore. I know what the disease wants.
> Ronnie: What does the disease want?
> Seth: It wants to turn me into something else. I'm becoming something that never existed before. I'm becoming . . . Brundlefly. Don't you think that's worth a Nobel Prize or two?[44]

As previously mentioned, there are two scenes, like bookends on either side of his initial self-transmission, in which Brundle is cinematically narrated on a screen within a screen. The second of these scenes occurs just after the self-naming of the Brundlefly. In front of Ronnie's video camera for the benefit of Stathis, Brundlefly performs a kind of Mr. Wizard science skit on the self-referential topic, "how does Brundlefly eat?" With Stathis as the internal viewer of this revolting vignette, the former Seth and his vomit drop are now embedded to the third degree, multiply framed within the main cinematic frame by video camera display playing on the television monitor in Ronnie's apartment. As with the first scene of cinematic embedding, Cronenberg aligns the depth of narrative level with the depth of his main character's psychic and bodily abasement.

In the discussion of narrative vitalism in the previous chapter, we noted the formal resonance between narrative and reproductive embedding, between stories within stories and gestating fetuses within pregnant mothers. The similarity of Seth's telepods to maternal wombs has often been noticed.[45] This connection is compounded by the traditional correlation between metamorphic narratives, maternal transformations, and related family matters—in particular, the proverbial link between incest and monstrosity. The bodily transformations of maturation, pregnancy, and delivery in all sexually reproductive living systems map out a pervasive biological and natural subtext of metamorphic fantasies.[46] Cronenberg's *Fly* couples mechanical, social, and biological systems into a fantastic fusion of informatic duplication, narrative embedding, and sexual reproduction.

Just after Stathis views her video tape of Brundlefly at lunch, Ronnie reveals that she is pregnant by it/him. With this turn of events, the metamorphic themes and narrative forms of *The Fly* intertwine and come full circle. The virtual destination of this tale's forms of embedding is the fusion of the media technology of teleportation with the maternal function of sexual reproduction. Seth in his transmission pod is already a posthuman fetus in a technological womb; Ronnie's troubling pregnancy doubles and confirms this reproductive frame. The informatic transmissions from pod to pod reenact genetic transmissions from womb to womb, in this case making instantaneously spectacular the potential for copying errors in the fantastic fusions of acquired genomes, not to mention the normal recombination of genetic contributions from the meiotic forms of parental sex cells. In allegorical parallel with Seth's technological rebirth—the confusion of transmission that delivers the metamorphosing Brundlefly—this narrative now embeds itself with a scene of monstrous miscarriage.

Without marking any shift in diegetic level, the cinematic narration cuts immediately from the scene of Ronnie's confession of pregnancy to one of her arrival with Stathis at a hospital. To complete by surgical abortion, so it would seem, a partial miscarriage of the fetus she had conceived with Seth after his genetic fusion with the fly, Ronnie is wheelchaired into an operating room, where she delivers from the cryptopod of her womb a hideous slimy footlong squirming grub, either the premature fetus or full-formed infant monster sired by the Brundlefly. As she screams in terror, the scene immediately breaks, redistributing the shock of this horrific delivery back to the main narrative. Only now the cinematic narrator marks the embedded frame of that episode by cutting to Ronnie thrashing awake from a nightmare and curling into fetal position. The ontological downshift back to the main storyworld supplies a cheap and nasty narrative lurch, in line with the vicious or backhanded recursivity of the text as a whole.

In addition to the earlier cinematic framings of screens within screens, this scene endows the film with an ultimately if not initially explicit moment of diegetic or narrative embedding by placing an image of abortive reproduction into a dream fold of the main storyworld, and then aborting it. Ronnie's dream of witnessing herself deliver the monstrous grub is the direct counterpart of the scene of Hélène's witness of André as monstrous man-fly, and in Neumann's version, the fly-headed André's fly-eyed countershot of Hélène. As internal focalizors, both the wife and the girlfriend are victimized by visions of "scientific" monstrosity, spectacles the most monstrous precisely as beheld by these intimate beholders. And both scenes of visual assault are marked by rude twists put on forms of narrative embedding.

The Brundlefly Project

As a discrete short episode embedded within the main narrative, Ronnie's abortion dream is also embedded within the longer episode of the Brundlefly Project, the unfolding of which draws Cronenberg's version of the *Fly* narrative to a close. But at the same time that the narrative presses toward the posthuman emergence of "something that never existed before," it is also retrofitted with a eugenic scenario that reinstates the theme of genetic purity in the midst of the genetic fusions. Echoing classical metamorphic romances seeking narrative resolution in the return of the lost human form, as the Brundlefly is increasingly expressed in the creature that Seth is becoming, the Brundlefly Project seeks to thwart or control that process—to "refine" it, limiting its posthuman consequences by further infusions of "pure human subjects." This mishmash of motives wonderfully twists out the ontological complications of a creature that is literally a two-sided form and of two minds about its future prospects. As Brundlefly pecks at the keyboard with metamorphosing hands bearing fused digits, the computer picks up the narration by reading out the following display:

THE BRUNDLEFLY PROJECT

PROBLEM: TO REFINE FUSION PROGRAM

GOAL: TO DECREASE TO A MINIMUM

THE PERCENTAGE OF FLY

IN BRUNDLEFLY

> SOLUTION: THE FUSION BY GENE-SPLICING

OF BRUNDLEFLY INTO ONE OR

MORE PURE HUMAN SUBJECTS[47]

But at this very moment, the attempt to counter the posthuman is thwarted by a cybernetic twist on the posthuman turn in metamorphic tales: that aphasic moment when the metamorph first tries to speak but can only, like Apuleius's Lucius, bray like an ass or, like Kafka's Gregor, chirp like an insect. Once Brundlefly absorbs the readout above, it voices a further oral command: "I want a disk—give me preliminary information."[48] But the computer's voice recognition program marks and narrates the precise moment when the process of organic posthumanization proceeds across the line of humanity into unrecognizable hybridity:

ERROR—PATTERN MISMATCH

VOICE NOT RECOGNIZED

VOICE NOT RECOGNIZED

VOICE NOT RECOGNIZED . . .[49]

In H. G. Wells's *The Time Machine*, accompanied by his devolved posthuman companion Weena, the Time Traveler comes across a ruined museum wherein the world of that future had forgotten a past containing his former present. Cronenberg famously places into his contemporary *Fly* the Brundlefly Museum of Natural History, wherein are collected the bodily relics of its/his human past. Ronnie returns to tell the being she still regards as Seth about her pregnancy and decision to terminate it, but before she can deliver that message Brundlefly drives her off with a warning about its/ his irreparable if temporary bifurcation into a two-sided being, oscillating between and thus "representing" both male human and male insect subject positions:

> Seth: Have you ever heard of insect politics? Neither have I. Insects don't have politics. . . . We can't trust the insect. I'd like to become the first . . . insect politician.
> Ronnie: I don't know what you're trying to say.
> Seth: I'm saying, I'm an insect who dreamt he was a man and loved it. But now the dream is over, and the insect is awake.[50]

The contrast between Brundlefly's rewording of the sense of being an "insect politician" and that dialogue's probable source in Taoist scripture offers a striking comparison between the cybernetic present and the mythological past, Western rationality and Eastern religion.

> Once upon a time, I, Chuang Tzu, dreamt I was a butterfly, fluttering hither and thither, to all intents and purposes a butterfly. I was conscious only of following my fancies as a butterfly, and was unconscious of my individuality as a man. Suddenly I awaked, and there I lay, myself again. Now I do not know whether I was then a man dreaming I was a butterfly, or whether I am now a butterfly dreaming I am a man. Between a man and a butterfly there is necessarily a barrier. The transition is called metempsychosis.[51]

The classical Taoist sage sees the species barrier between man and butterfly as transcendable only after death with the transference of the soul from one to another body—a motive for the narration of metamorphosis going back to its archaic mythological sources. But given that barrier, Chuang Tzu's focus is on the undecidability of the oscillation, that is, the paradox presented by the dream life that lies embedded within the mind with a reality of its own putting one's waking reality into question. In contrast, Brundlefly is the product of an immediate living–mechanical transcendence of the species barrier—that is, it/he is a modern media metamorph and not a mythical serial metempsychotic. In the fly mode of *its* subject-oscillations, moreover, it denies the undecidability of the question. As the narrator of Kafka's *Metamorphosis* had put it regarding Gregor's uncanny awakening:

"It was no dream."[52] So it/he gives Ronnie fair warning that man and fly will cohabit this metamorph only until the awakened fly takes full control: "I'll hurt you if you stay."[53]

And despite Ronnie's taking the hint and running away to procure an abortion for real, it is not the fly but the Brundle in Brundlefly who bursts into the operating room and carries Ronnie back to the lab. Their baby "might be all that's left of the real me," without which the Brundlefly Project will fall short of its "purest" goal, "TO DECREASE TO A MINIMUM THE PERCENTAGE OF FLY IN BRUNDLEFLY." Technoscientific to the bitter end, Brundlefly remains, as Seth had been, the object of its/his own project in posthuman self-fashioning. As Bruno Latour writes, when technological designs begin on the drawing board, "there is no distinction between projects and objects. . . . Here we're in the realm of signs, language, texts"; Brundlefly's attempt to refine its/his own posthuman constitution fictively reflects the technological innovator's actual quest to "translate" or "negotiate with" both the human and the nonhuman components of the design: "the innovator has to count on assemblages of things that often have the same uncertain nature as groups of people. . . . You have to get a whole list of things interested *in the project.*"[54]

In the climactic scene that follows, Brundlefly negotiates with humans who are decidedly not "interested in the project." Ronnie refuses to abort her quest for an abortion, and Stathis shows up at the lab with a skeet-shooting rifle. It will also turn out that the "assemblages of things" involved—the teleporter and its rewired pods—will "have the same uncertain nature." This terminal episode further entrenches the traditional themes of metamorphic embedding at both the bodily and narrative levels within the cybernetic realm of machinic couplings with organic forms. As a monstrous organic being morphed by the teleporter, accordingly the Brundlefly has now remorphed the teleporter. Initially conceived as a linear transportation device moving goods from pod 1 to pod 2, by splicing in the previously decommissioned "clunky" prototype pod, Brundlefly reconfigures the teleporter precisely as it had demonstrated itself to be in its own instance: a nontrivial, recursive, or recombinant gene-splicing device, now retrofitted for the "creative" merging of the contents of pod 1 and pod 2 into a composite arriving at pod 3. With Stathis once again as internal focalizor, the computer narrates:

GENE-SPLICING METHODOLOGY

HARDWARE:

TELEPOD 1: TRANSMITTER OF SUBJECT A

TELEPOD 2: TRANSMITTER OF SUBJECT B

TELEPOD 3: RECEIVER OF GENETICALLY-FUSED

   A-B COMBINATION SUBJECT

After putting Stathis out of commission with vomit drop, Brundlefly accedes to Ronnie's plea to spare his life to reopen a final round of negotiations with her: "Help me be human . . . more human than I am alone." The consummation of the Brundlefly Project marks the systemic closure of the human itself in its inhospitality to the nonhuman, its doggedly in-turning self-involvement. But by now, insect politics have demented Brundlefly's traditional humanistic family values: "I go there, you go there, we come apart and come together—there: you, me, and the baby . . . together. . . . We'll be the ultimate family."[55] One can only expect from the climactic three-way fusion of Brundlefly, Ronnie, and their fetus a monstrosity worthy of its fetishistic precursor, Langelaan's climactic three-way merger of André, the fly, and the cat Dandelo.

But we are spared that particular organic vision in favor of a truly cybernetic climax—Brundlefly's consummation in a terminal cyborgian fiasco. To set it up, Ronnie's resistance to its final negotiation precipitates the final collapse of Brundlefly's two-sided physical form: the "space bug" or monster fly—the imago or adult stage of the metamorphic Brundlefly—hatches entirely out of the "veil of the flesh," the tattered chrysalis of Seth Brundle's human body.[56] It is a moment both visually horrific and conceptually spectacular because it renders to view the phenotypic resolution of the organic process set into motion by the inadvertent genome splicing of Brundle and fly. Thus, while terrifying, it also produces a satisfaction in the narrative completion of a metamorphic plot. But Cronenberg is not done yet.

In other, affirmative stories of posthuman metamorphosis, such as Damon Knight's *Beyond the Barrier* or Octavia Butler's *Imago*, the arc of the plot does end here, with the achieved transcendence of the human. A being emerges into which the human has been absorbed, within which it is rendered, like the mitochondrion that was once a free-living bacterium, one of several components of a higher-order living system. Contemporary work in biological systems has strongly endorsed the view that natural evolution is driven not merely by random mutation but more importantly by the integration of separate genomes into viable consortiums. In this way *The Fly* actually has it half right: separate genomes can and do join forces, but when this happens, they are not just randomly scrambled together, as implied by the computer's report that Seth and the fly have been "FUSED AT THE MOLECULAR-GENETIC LEVEL."[57] Rather, in the accretion of genomes within a host organism, their given or preevolved structures are left largely intact but operationally coupled together to produce a higher-order transformation—a natural metamorphosis—of the living system at hand. Lynn Margulis, the biologist at the forefront of this neocybernetic refinement in evolutionary thought, puts it like this: "Analogous to improvements in computer technology, instead of starting from scratch to make all new modules again, the symbiosis idea is an interfacing of preexisting modules. Mergers result in the emergence of new and more complex beings."[58]

From this vantage we can see again the way that Cronenberg's *Fly* teeters on the conceptual divide between first- and second-order cybernetics, or rather, between the demonization and the productive unfolding of paradoxical recursions. But in the final resolution of this recursive narrative arc, as the composite teleporter feeds the output of subsystems back into its input, the organic and the mechanical systems at hand are not merged into a consortium of semiautonomous modules, but scrambled together into a horrific hash, an outcome neither literally nor fictively necessary. In the unfolding of *Fly* mythology up to this point, the posthuman remains ultimately foreclosed and the monster must die. Thus we are given a choice between two equally exquisite nonviable options: instead of the "ultimate family" of Brundlefly's nostalgic organic desire, we get the terminal fusion of the organic space bug with the mechanical telepod itself: "FUSION OF BRUNDLEFLY AND TELEPOD SUCCESSFUL."[59]

This narrative closes with the consummation of Brundlefly's death spiral within its/his own cybernetic loop: the Seth-sentience left within the scrambled creature makes its/his final self-referential gesture, crawling out of the receiver pod to aim the skeet rifle in Ronnie's hands toward its/his brains and effectively imploring her to render its/his euthanasia. But this conclusion also marks the family-romantic nostalgia that folds Cronenberg's *Fly* back upon Neumann's version of the conclusion, as the surviving mother and son go off into the sunset with Vincent Price as the father/surrogate. This *Fly* consumes itself by its own death drive to resolve its productive detour in the restoration of its prior narrative conditions. Thus we get the message that posthuman metamorphosis is inevitably a death trip, as the "sex appeal of the inorganic" trumps—for this round of the mythic cycle at least—the living systemic imperative of autopoietic continuation.[60] But the repetition compulsion works both ways, and one anticipates that media conditions within the social subsystems responsible for *The Fly* will, at some future moment, revive this *Fly* once again.

SIX

Posthuman Viability

"Hard SF" is distinguished from its fellow genres by the relative verisimilitude of its scientific representations. Modern stories of bodily metamorphosis adapt themselves to this generic pressure at the level of their biological motivations. Both the plausibility and the viability of the depicted transformations can be measured against the life-scientific knowledge and theory standing across the border from the storyworld, at the time of composition and thereafter. But the meanings circulated by stories of metamorphosis also oscillate in the space between psychic constructions and social discourses—and from our particular perspective, that space can be plotted through a neocybernetic understanding of the interpenetration of psychic and social systems.

Like the protagonist of *The Fly*—a catastrophic divergence from the biological norm with next to no hope of establishing biological, psychic, or social viability—the narrative construction of a modern metamorph can take the form of an individual singularity. Such depictions are in line with

mythical and classical literary precursors. But modern metamorphs sometimes take an alternative form: as with the Beast People in *The Island of Dr.
Moreau*, they emerge within an entire population, which must then grapple
with a general and distributed corporeal transformation. In keeping with
the modern understanding of evolutionary species, only the latter scenario
offers plausible grounds for the depiction of a viable transformation. But
then, it is only with the emergence of hard science fiction that the issue of
viability arises at all, alongside the issue of verisimilitude. Yet, whether viable or not, these modern metamorphoses—even in the nonscientific and
singular strain represented by a work like Kafka's *Metamorphosis*—
concentrate considerably increased attention on the desire of metamorphs
to achieve social inclusion, and generate pathos by chronicling the wreck of
such desires. Just before the collapse of the Brundlefly Project, Seth exclaims, "we'll be the perfect family."

For instance, in *The Island of Dr. Moreau*, the topical verisimilitude of
Wells's science fiction lies in its knowledgeable ("hard") depiction and
intermingling of vivisection and devolution, those hot-button scientific-
ethical issues of the Victorian fin-de-siècle. The fantasy of forcing humanization upon animals through surgical alterations pitches the ensuing failure
of those experiments—the return to bestiality—as an allegory of human
devolution. The apocalyptic cascade of reverse metamorphoses that concludes the novella dooms the transformations of its hapless metamorphs to
nonviability. Nevertheless, the most prescient stroke of the narrative is the
way that the Beast People suffer *en masse* their promotion to and relapse
from a marginal humanity. They are decisively developed in their struggle,
in systems-theoretical terms, to stabilize their psychic and social autopoiesis—their metabiotic and not simply their biological self-maintenance.

As we discussed in chapter 2, the cruelest twist of this narrative lies in
the way that Dr. Moreau ultimately forecloses the Beast Peoples' desires—
posed in the rhetorical question of their pitiful chant, "are we not
men?"—by his persuasion to the contrary. Had he held faith with his own
creations, as it were—had he truly lived up to his own Promethean aspirations—that alternative persuasion might have tipped the scales in the other
direction. Put another way, had he constructed his own perceptions of his
Promethean accomplishments according to a different, *post*human set of assumptions, one imagines he might have been a more successful constructor
of "uplifted" creatures.

But the possibility of this speculation simply reminds us how the limits of Wells's own conceptuality—and hence that of his dark avatars Dr. Moreau and Prendick—were drawn by a theorization of evolution that remained conflicted over the clash between, on the one hand, Darwin's progressive intimation of the nonessentiality of the concept of species, and the general and regressive anxiety to salvage the notion of the sanctity and inviolability of the human from that recognition. Although the transformations of *The Fly* update their technologies of scientific metamorphosis, the fabulations that go by that name, so far at least, have not escaped the general pull of humanistic anachronism on evolutionary ideas.

As often as not, the rise and proliferation of cybernetic concepts and images have run in tandem with older evolutionary ideas pitting the humanist image of the human against the monstrosity of the posthuman. For instance, for all of its gritty future-technological ambience, male-authored cyberpunk is not especially distinguished by its transcendence of patriarchal nostalgias. Take Bruce Sterling's *Schismatrix*: while it makes "Posthumanism" an explicit theme, still, in the strife of its Shapers and Mechanists it retails an all-too-human oppositionalism between living and nonliving systems, diminishing the provenance of cybernetics to abiotic matters.[1] "'This struggle is about Life,'" says the character Constantine: "'the Mechs . . . will be cybernetic, not living flesh. That's a dead end, because there's no will behind it. No imperatives. Only programming. No imagination.'"[2] Sterling's narrative struggles to complicate such polar scenarios, but, in its denouement, the bioengineered ("shaped") metamorphic body shown to the defeated Constantine by his friend and rival Lindsay—displayed like a retirement estate in a promotional brochure—has a telling flaw related to its commodity status as a posthuman lifestyle option:

> The center pages showed an Angel's portrait: an aquatic posthuman. The skin was smooth and black and slick. The legs and pelvic girdle were gone; the spine extended to long muscular flukes. Scarlet gills trailed from the neck. The ribcage was black openwork, gushing white feathery nets packed with symbiotic bacteria. . . . The lidless eyes were huge and the skull has been rebuilt to accommodate them.
>
> Constantine moved the brochure before his eyes, struggling to focus. "Very elegant," he said at last. "No intestines."
>
> "Yes. The white nets filter sulfur for bacteria. Each angel is self-sufficient, drawing life, warmth, everything from the water."

"I see," Constantine said. "Community with anarchy. . . . Do they speak?"
Lindsay leaned forward, pointing to the phosphorescent lights. "They glow."[3]

Their light-language notwithstanding, these bodies designed for warm
ocean life beneath the icy crust on Jupiter's moon Europa achieve a satur-
nine perfection that is anything but posthuman—the rugged Western ideal
of being "self-sufficient." Here viability is keyed more to "anarchic"
detachment than to the "communal" sociality intimated by the Gaian symbiosis of Angels and bacteria. But some narratives of posthuman metamorphosis do escape the gravity of retro humanist constructions, seizing on the
metamorphic potential of evolutionary and cybernetic motifs to rework the
cultural typifications of mainstream science fiction and cyberpunk in the
direction of posthuman viability. Certain female authors of hard science
fiction in particular, such as Ursula Le Guin, Joan Slonczewski, Marge
Peircy, and Octavia Butler seem especially adept at developing more profound treatments of the question: what forms best imagine the potential
viability of posthuman evolution?

Octavia Butler's Metamorphoses

Octavia E. Butler (1947–2006) was the foremost African-American feminist
author of science fiction of her time. Spanning four decades and fifteen
volumes, Butler's fictions reside in allegorical landscapes just to one side or
over the horizon of the historical world. Placing her fiction on these alternative planes, she approached a range of social issues painful to probe in their
stark historical or personal reality. Butler chose only once to confront
American racial history head-on. Her acclaimed novel *Kindred* mysteriously
transfers Dana, a post-civil rights movement African-American woman in
an interracial marriage, back and forth from horrific adventures in antebel-
lum Maryland. Yet this one mainstream book with a time-travel twist from
her earlier career indicates the default decoding of all the others—slavery,
racial conflict, cultural domination and social oppression, and class and gender inequity, as these have plagued Western civilization in general and
American society in particular.

Butler's work is most often discussed with regard to these sociopolitical
matters of American racial history in particular and of sex and gender

relations in general.[4] Her writings allude to these historical contexts with an uncanny sense for the ironies that bind master and slave, oppressor and victim, alien and aborigine, together in a common destiny. Tilting her narratives out of realism to alleviate the sting of immediate recognition, she induces a bittersweet reading experience that adds the bonus of literary pleasure to the confrontation with cultural demons. Butler forecloses any easy ways out of confronting the dilemmas of survival and coexistence—out of having to rethink old destructive habits. But she cushions anticipations of cataclysm with images of posthuman possibility. Probing the fissures of society, race, and species, Butler's fictions find rare moments of improbable accord. Played out through the transformations of self-conceptions in response to the metamorphoses of bodies and environments, her theme is the interpenetration of individual, social, and planetary changes.

Our focus here is on the scientific and technological details of posthuman metamorphosis in Butler's texts, and in particular on the remarkable Xenogenesis trilogy—*Dawn* (1987), *Adulthood Rites* (1988), and *Imago* (1989). From this angle, her fiction is a meditation on evolution: the worlds she invented are populated by individuals and groups placed under evolutionary pressures. Some are human, some are mutant, and some are extraterrestrial. Still others are the children of unions between humans and mutants, or humans and extraterrestrials. Sometimes alone but typically in small groups, families or packs, her characters have to carry the weight of evolutionary developments—the specter of deliberate or inadvertent extinction, the divergence of natural or constructed variations, the selection of hybrids or of the bearers of mutations, and the responsibility for cultivating new evolutionary vectors through which a future can be purchased at the price of irrecoverable loss and irrevocable change. In Butler's worlds the only thing for it is to hold fast to alien couplings that maintain life and the possibility of survival within altered environments.

I. THE HARAWAY CONNECTION

Our glance at Butler's Parable novels will be brief: *Parable of the Sower* (1993) and *Parable of the Talents* (1998) retract science-fictional speculation to the bare limit of the genre.[5] Metamorphic fabulation is set aside: these near-future dystopias center images of transformation on the unfolding of

Earthseed, Lauren Olamina's campaign to make her prophecy of space colonization the basis for a self-perpetuating, extraplanetary social system. But the fabula itself never gets off Earth. Lauren's secular religion of self-change in face of fundamentalist intolerance and social dispossession is shaped through the construction of a self-reproducing body of scriptures. Like autopoietic social systems in Luhmann's description, the spirit of Earthseed is constructed as observant but without consciousness: "Olamina believes in a god . . . not consciously aware of her—or of anything. . . . 'God is Change,' she says and means it. Some of the faces of her god are biological evolution, chaos theory, relativity theory, the uncertainty principle, and, of course, the second law of thermodynamics."[6] As in the passage above, in these texts scientific concepts serve as a backdrop and as components of a mindset, but they are not decisively dramatized.

But Butler's earlier novels virtually enact the supernatural powers and natural sciences they discuss and ring changes on change itself with narratives of physiological and psychic metamorphosis. For instance, her first series—the Patternist novels—develops both pro- and retrospectively through the generations of a mutant population dispersed among "normal" humans.[7] This series could be called hardcore fantasy punk—it all but touches on the cyberpunk genre it slightly predates, except that the Pattern that links mutant minds across distances is not a media-technological cyberspace but some sort of organic telepathy field. From the literary warm-up of the Patternist novels to the abiding achievement of the Xenogenesis trilogy, Butler' work remains rooted in social allegories of race, gender, and power, but shifts from supernatural fantasy to mainline hard science fiction. In her widely circulated "Cyborg Manifesto" and elsewhere, Donna Haraway's seminal critical attentions to Octavia Butler captured these shifts from sorcery and slavery to apocalyptic genetics, setting down the lines for a "cyborg" approach to Butler's writings that continues unabated.[8]

> Octavia Butler writes of an African sorceress pitting her powers of transformation against the genetic manipulations of her rival (*Wild Seed*), of time warps that bring a modern US black woman into slavery where her actions in relation to her white master-ancestor determine the possibility of her own birth (*Kindred*). . . . In *Dawn* (1987), the first installment of a series called Xenogenesis, Butler tells the story of Lilith Iyapo, whose personal name recalls Adam's first and repudiated wife and whose family name marks her status as the widow of the son of Nigerian immigrants to the US. . . . Lilith mediates the transformation of

humanity through genetic exchange with extraterrestrial lovers/rescuers/destroyers/genetic engineers, who reform earth's habitats after the nuclear holocaust and coerce surviving humans into intimate fusion with them. It is a novel that interrogates reproductive, linguistic, and nuclear politics in a mythic field structured by late twentieth-century race and gender.[9]

Haraway's *Modest_Witness@Second_Millennium* elaborates the technoscientific transgresson of previous ontological categories by detailing the new mergers that result from them: "imploded germinal entities, densely packed condensations of worlds, shocked into being from the force of the implosion of the natural and the artificial, nature and culture, subject and object, machine and organic body, money and lives, narrative and reality."[10] This ambitious text develops both the deconstructive and participatory dimensions of cyborg discourse as a mode of cultural–allegorical *figuration* with deep roots in Western scriptural hermeneutics: "figuration [produces] secular technoscientific salvation stories full of promise. . . . Figuration in technoscientific texts is often simultaneously apocalyptic and comedic."[11] Haraway and Butler work from opposite ends of the creative spectrum toward a vision of contemporary secular mythology adequate to neocybernetic reality.

In the Xenogenesis trilogy, the capture and transformation of humanity by the alien Oankali, "natural genetic engineers,"[12] enacts the cultural repercussions of a technologized bioscience that has shredded any notion of inviolable natural or cultural realities withstanding its transformative doctrines and techniques. As Haraway's *Modest_Witness* affirms on a stack of technoscientific evidence: "universal nature itself is fully artifactual."[13] "Transported" by the real transuranic and transgenetic technologies that produce postnatural elements such as plutonium and organisms such as tomatoes with flounder genes, "terran chemical and biological kinship get realigned to include the extraterrestrial and the alien."[14]

Although *Modest_Witness* does not extend Haraway's discussions of the Xenogenesis trilogy, it may be said virtually to incorporate that work of fiction: a fictive world in which human beings enter procreative alliances with metamorphic aliens who "grow" machines *is* our world of "crossovers, mixing, boundary-transgressions"[15]—a world, that is, of universal material and genetic constructivism. Through alien bodily organs the intergalactic Oankali not only neutralize the morbidity and retrieve the creative potential of cancer, but also organomorphose the mechanical, turning heavy metal into tough cellular hide. As Lilith explains to a fellow human who has recently entered the "trade" relationship with the Oankali, regarding the Lo entity implanted in the Amazonian earth beneath and all around them: "It's an Oankali construct. Actually, it's a kind of larval version of the ship. A neotenic larva. . . . Its inclination . . . is to become a closed system. A ship. We can't let it do that here. We still have a lot of growing to do ourselves."[16] In this passage the "closed system" held in abeyance is properly neocybernetic—the autopoietic spacegoing biosphere genetically constructed by Oankali forebears, within which the construct posthumans will eventually depart the planet.

But the posthuman still has deep roots in the premodern. In the chapter of *Modest_Witness* devoted explicitly to racial matters, Haraway connects the figure of the cyborg to that of the vampire, a boundary-transgressor between life and death, purity and infection.[17] Butler's fiction, with its powerful racial undertones, is also well-bitten by vampirism. But when the Oankali extrude microfibrous sensory organs to pierce human flesh—to inject organic drugs, probe cells, absorb genes, and alter bodies—their parasitism is mutualistic rather than predatory. Lilith, the matriarch of Xenogenesis, is the first traitor to the human, who takes posthumanity upon her when she accepts her Oankali companion Nikanj's offer of bodily upgrading and submits herself to the hypodermic prick of its sensory tentacles. The subsequent genetic merger of human and Oankali leads to construct offspring who metamorphose further with each generation: although Lilith's child Akin looks human at birth, it has an Oankali tongue: " 'Regular little vampire,' [Akin] heard . . . before he was lost in the taste of her."[18]

2. VAMPIRES AND METAMORPHS

Only in Butler's last novel, *Fledgling* (2005), does the vampirism refracted in the earlier fiction become explicit: the novel centers on the character Shori, a genetically modified African-American vampire. The Pattern that gives the name to the Patternist series is a telepathic web of psychic connections reminiscent of the vampiric clairvoyance by which Mina Harker becomes a medium or spirit receiver–transmitter for the Count in Bram Stoker's *Dracula*. The patriarch of the Patternists is Doro, an incorporeal spirit that lives in near immortally by leaping from body to body: the person occupying his next one is erased, the prior body is vacated and discarded and dies on the spot. For several thousand years he has been breeding children, taking the bodies of some when he needs a fresh one and seeking eugenic improvements of mutant capacities for the rest. Doro's lethal embraces are more blatantly Transylvanian than the Oankali's sensory-tentacle lovebites.

The figure of the vampire represents a kind of Gothic, anachronistic posthumanity. The elite among them, such as Stoker's Count, may be metamorphic, but most are only marginally so, as their bodily differences are largely concealed and held away from the sunlight. As metamorphs go, they remain liminal, hovering at the boundary between night and day, mortality and immortality. While in the vampiric vein of the Patternist series, Butler disposes of the more shadowy elements of the figure, setting her mutants forth into battle for viability and progeny with nontraditional impediments generally psychic rather than corporeal. Even in *Fledgling*, the genetically engineered alterations that place Shori and her immediate relatives in mortal danger from their less progressive brethren also "normalize" her, render her relatively impervious to the effects of sunlight. Nonetheless, even with the biotechnological updating that Butler performs in *Fledgling*, the vampirism evoked remains a form of posthuman transformation contingent upon premodern convictions about blood as the essence of life and racial identity. Similarly, the metamorphic imagination that plays in various ways through the Patternist series operates according to classical patterns of instantaneous magic rather than contemporary tropes of cybernetic or genetic control.

As the posthuman mutations of the Patternists go, Doro is demonic, a tireless vampiric force of dispossession, while his female counterpart in *Wild Seed*, Anyanwu, is daemonic, a witch with metamorphic healing powers.[19] Derived from African legends of shape-shifters, Anyanwu undergoes spectacular cross-species transformations into birds, felines, and dolphins: "I took animal shapes to frighten my people when they wanted to kill me," Anyanwu tells Doro during their first encounter while still in Africa:

> "I became a leopard and spat at them. They believe in such things, but they do not like to see them proved. Then I became a sacred python, and no one dared to harm me. The python shape brought me luck. We were needing rain then to save the yam crop, and while I was a python, the rains came. The people decided my magic was good and it took them a long time to want to kill me again." She was becoming a small, well-muscled man as she spoke.[20]

Doro takes Anyanwu for a mate and carries her along when he departs Africa by slave ship for the New World, which he has already reconnoitered as a land fit for the extension of his mutant empire. As Doro forces Anyanwu into his millennial breeding program, holding her fast by threats of violence to their children, the narrative follows this same pattern of metamorphosis as a fight-or-flight mode of self-defense. Much as she longs to wander off on metamorphic escapades, Anyanwu keeps coming back to her abusive home. This dynamic is perfectly captured when the slave ship nears the coast of New York, in a scene of reverse metamorphosis after she returns from a brief interlude as a dolphin:

> She began to grow legs. . . . And her flippers began to look more like arms. Her neck, her entire body, grew slender again and her tiny excellent dolphin ears enlarged to become less efficient human ears. Her nose migrated back to her face and she absorbed her beak, her tail, and her fin. There were internal changes those watching could not be aware of. And her grey skin changed color and texture. That change caused her to begin thinking about what she might do to herself if someday she decided to vanish into this land of white people she was approaching. She would have to do some experimenting later. It was always useful to be able to camouflage oneself.[21]

In contrast, metamorphosis in the Xenogenesis trilogy is not a singular adaptive mutation but a species-specific evolutionary acquisition yielding xenogenic powers of biological manipulation compatible with human genetics. Biologist Joan Slonczewski, herself an accomplished author of science fiction novels, has written how Butler's narrative "is presented in terms remarkably consistent with modern molecular biology, even predicting developments that have occurred since the novels were written. . . . Unlike the vast majority of alien abduction tales, *Dawn* actually presents a biologically plausible explanation for why the Oankali need to interbreed with humans."[22] In the Xenogenesis trilogy, Butler's remarkable consistency with modern bioscience, as we will see, is well earned and puts her depictions of bodily metamorphosis on an entirely different footing than that of the Patternist novels. She built up and extrapolated her posthuman biological scenarios there from specific popular-scientific sources that were not only reliable but visionary in their own right. Several of these key discourses were just coming into broad notice as Butler began to incubate this narrative in the mid 1980s. Now with two decades of hindsight one can affirm how profoundly sound her fictive intuitions were in response to the scientific works in question.

The Xenogenesis Trilogy

I. THE MARGULIS AND SAGAN CONNECTION

It has not gone unnoticed that Butler's Xenogenesis trilogy strongly resonates with the popular writings of microbiologist Lynn Margulis and science journalist Dorion Sagan. Margulis and Sagan have published a series
of volumes—*Microcosmos: Four Billion Years of Evolution from Our Microbial
Ancestors* (1986), *Origins of Sex: Three Billion Years of Genetic Recombination*
(1986), *Mystery Dance: On the Evolution of Human Sexuality* (1991), *What is
Life?* (1995), *What is Sex?* (1997), and *Acquiring Genomes: A Theory of the
Origins of Species* (2002)—developing the considerable implications of a biological theory that Margulis, against a now-vanquished opposition, salvaged
from disrepute and retooled through methods of genetic analysis developed
in molecular biology. Initially called SET (Serial Endosymbiosis Theory),
it is now usually referred to as *symbiogenesis*—the evolution of new life forms
out of the symbiotic merger of preexisting life forms. The range of the
theory itself has developed since the 1970s, when the focus on "serial endosymbiosis" centered primarily on determining the sequence in which bacterial symbionts took up residence within host bacteria to produce and
augment the consortium whose evolutionary viability stabilized as the nucleated or eukaryotic cell. Symbiogenesis applies to all theories citing sym-
bioses that "cross kingdoms" in various ways between bacteria, protoctists
(such as algae), fungi, animals, and plants to explain further events of evolutionary speciation.

But do we know for certain that Butler actually read anything of Margulis
and Sagan's while the trilogy was being conceived and written? In "Dialogic
Origins and Alien Identities in Butler's Xenogenesis," Cathy Peppers has
briefly developed the Xenogenesis–symbiogenesis connection, supporting
the link by reference to a personal communication in 1994.[23] In the head
note to her endnotes, Peppers wrote: "Thanks also to Octavia Butler, whose
assessment of Wilson and Margulis in a quick hallway conversation encour-
aged me to keep questioning the 'big guns' of Big Science."[24] The "big
gun," relative to Margulis, would be the entomologist and sociobiologist
E. O. Wilson, recently termed "the alpha male of Literary Darwinism."[25]
In the body of the article, Peppers cites from a 1991 anthology's excerpt of
Margulis and Sagan's *Microcosmos*, culminating in their remark that " 'guests

and prisoners can be the same thing, and the deadliest enemies can be indispensable to survival'"; Peppers comments: "It's hard to think of a better representation of the relationship between the Oankali and the humans in Xenogenesis. . . . In choosing to privilege Margulis' symbiotic story of origins over the traditional Darwinian one, Butler is able to expose and contest the eugenic aspirations driving the latter."[26]

Peppers documents that by the mid 1990's, Butler knew the work of Margulis enough to contrast it with that of Wilson. In 2005 I e-mailed Butler to enquire about the information in Peppers' essay and received the following response:

> I have read Lynn Margulis, James Lovelock (Gaia), and E. O. Wilson. I was more interested in Wilson's writing about social insects than his sociobiology theory. But the first book of his that I read was *On Human Nature*. I remember having a great time reading it and arguing with it. Sometimes I get the best ideas from books I don't agree with.
>
> Lynn Margulis directly affected what I did in the Xenogenesis novels. . . . I discovered Margulis and Lovelock through the PBS television show, NOVA. The show dealt with Lovelock's Gaia theory and Margulis spoke strongly in favor of it. I was fascinated by both scientists. I sent for the transcript. Then I went first to the library and then to the bookstore to get books by both. For starters, I found *Microcosmos* and *Gaia: A New Look at Life on Earth*. I went on being fascinated and doing some arguing with them. I've been reading what I could find of both ever since.[27]

The mediation of the NOVA documentary gives some insight into Butler's working methods. Titled *Gaia: Goddess of the Earth*, featuring numerous samples of whole earth footage showing our world and its biosphere seen from space (i.e., the Oankali view of things), engaging vignettes of Lovelock and Margulis in defense of the theory, with expert critiques from Stephen H. Schneider and Richard Dawkins, it aired on Tuesday, January 28, 1986— the day of the Challenger disaster. This circumstance surely diminished its total audience, but may have also nailed its impact that much more powerfully into Butler's creative meditations for an extended story predicated on a human catastrophe (the nuclear holocaust that occurs offstage). It may have suggested, for instance, the ironic reversal, which centers Butler's entire fabula, by which mishap-prone humans relinquish the role of scientific observers in favor of a race of space-faring, gene-trading explorers who have gotten themselves right with life.

The Challenger disaster is all the more ironically poised against NOVA's *Gaia: Goddess of the Earth* because of the latter's photographic and cinematic images drawn from NASA cameras during missions putting Americans into orbit around the Earth and the Moon. The space program that enabled these reverse focalizations of the biosphere—all those iconic tableaus of Earthrise over the lunar landscape—linking the Whole Earth sensibility to the unfolding of Gaia theory and its biological, social, and philosophical spin-offs, was also the primary source of funding for Lynn Margulis's work on SET and symbiogenesis.[28] But when one opens up the work Butler turned to after viewing the Gaia documentary and reading its transcript, and which "directly affected what I did in the Xenogenesis novels"— Margulis and Sagan's 1986 volume *Microcosmos: Four Billion Years of Evolution from Our Microbial Ancestors*—one finds in that volume an even more trenchant reversal of perspective.[29]

In the preface to the paperback edition first released five years later, Margulis and Sagan reflected on the original rhetorical strategy of their *Microcosmos* narrative:

> Reversing the usual inflated view of humanity, we wrote of *Homo sapiens* as a kind of latter-day permutation in the ancient and ongoing evolution of the smallest, most ancient, and most chemically versatile inhabitants of the Earth, namely bacteria. . . . We outraged some with the implication that even nuclear war would not be a total apocalypse, since the hardy bacteria underlying life on a planetary scale would doubtless survive it. . . . We believe this formerly slighted perspective is a highly useful, even essential compensation required to balance the traditional anthropocentric view which flatters humanity in an unthinking, inappropriate way. Ultimately we may have overcompensated.[30]

But in terms of *Microcosmos'* contribution to Butler's literary achievement in the Xenogenesis trilogy, it is just as well that Margulis and Sagan engineered the bacterial counterperspective on prideful humanity as forcefully as they did. Butler fits her narrative directly to this frame.

The resonances between Margulis and Sagan's scientific argument and Butler's narrative enactment begin immediately in "Introduction: The Microcosm." Due to "a revolution . . . in the life sciences . . . established ideas . . . have exploded. . . . The simplest and most ancient organisms are not only the forebears and the present substrate of the earth's biota, but they are ready to expand and alter themselves and the rest of life, should we 'higher' organisms, be so foolish as to annihilate ourselves."[31] "A handful of people tried to commit humanicide," we read at the beginning of *Dawn*; "They had nearly succeeded."[32] *Microcosmos* continues: "the view of evolution as chronic bloody competition . . . dissolves before a new view of continual cooperation, strong interaction, and mutual dependence among life forms. Life did not take over the globe by combat, but by networking. Life forms multiplied and complexified by co-opting others, not just by killing them."[33] In Butler's fiction humanity's tooth-and-claw misconstruction of evolutionary dynamics feeds directly into the mode of its near annihilation. Moreover, the beings who rescue the human remnant exhibit social styles distinctly marked by cooperation, interaction, and interdependence. At the same time, their intentions are never to kill but rather to co-opt their captives through bacterial strategies of genetic recombination.

The tableau of bodily transformation in the Xenogenesis trilogy is centered on the extraterrestrial Oankali. A sizable population happen to be cruising by in their massive living ship when a full nuclear exchange on Earth draws their attention. Salvaging a few score of survivors along with what they can recover in samples of other terrestrial life forms, as the story begins they have been parked in extralunar orbit for several centuries, mopping up the devastated biosphere while, back on ship, they investigate the humans they have placed into suspended animation, learning their biology, language, and culture. As we saw in chapter 1, *Dawn* begins as the Oankali awaken Lilith Iyapo for good to begin training her for the lead role they have determined she will take in the "gene trade" to which they are bioconstitutionally devoted. As she will shortly be informed by her first mentor, Jdahya:

> "We trade the essence of ourselves. Our genetic material for yours."
> Lilith frowned, then shook her head. "How? I mean, you couldn't be talking about interbreeding."
> "Of course not." His tentacles smoothed. "We do what you would call genetic engineering . . . We do it naturally. We *must* do it. It renews us, enables us to survive as an evolving species instead of specializing ourselves into extinction or stagnation."[34]

In this cosmic allegory, the Oankali literally personify the primeval and perennial ways of the prokaryotes: both are "natural genetic engineers" and "gene traders." In *Microcosmos*, giving further details of the revolution in the life sciences based on "recently discovered dynamics of evolution,"[35] Margulis and Sagan note that, in addition to the understanding of how DNA works, "a second evolutionary dynamic is a sort of natural genetic engineering. . . . Prokaryotes routinely and rapidly transfer different bits of genetic material to other individuals."[36] Indeed, "by trading genes, bacterial populations are kept primed for their role in their particular environment and individual bacteria pass on their genetic heritage."[37]

Chapter 5 of *Microcomos*, "Sex and Worldwide Genetic Exchange," relates that "the first sort of sex to appear on our planet, the bacteria-style genetic transfer exploited by [human] genetic engineers . . . is simply the recombination of genes from more than one source."[38] Bacteria are not sexually dimorphic; they reproduce on their own by simple division involving only the replication of the DNA they already possess. The result of bacterial sex by lateral gene transfer is not reproduction but *metamorphosis* through the acquisition of new genes and the new metabolic capacities they carry. Now, Butler's Oankali are clearly sexual eukaryotes, but their processes of maturation and reproductive physiologies are still strongly marked with the ways of bacteria and other likely symbiogenetic participants. For instance, the individual bodies of Oankali and their cross-species constructs mature in a manner familiar to us from terrestrial animals, now considered to be of symbiogenetic origin, that pass through one or more larval stages on the way to corporeal adulthood.[39] Before metamorphosis, unless bioengineered otherwise (like Akin in *Adulthood Rites*), immature organisms hosting the Oankali organelle are *eka*, or sexually indeterminate. Although as children they are given gender assignments on the basis of what their immature predilections appear to be, the form of their *imago* or sexually mature adult stage is uncertain until that metamorphosis is under way.

Individuals that become adult males and females must be nursed through a single metamorphosis during which they grow new "sensory arms" and achieve mature sexual identity. But certain Oankali metamorphose into *ooloi*, individuals with a parasexual bodily form neither male nor female. These mature more slowly, and undergo two prolonged metamorphoses to reach adulthood. All Oankali and human–Oankali constructs possess a specialized organ, the "yashi," between their two hearts. Lilith's construct child Jodahs, the character-bound narrator of *Imago*, explains that "males and females used it to store and keep viable the cells of unfamiliar living things that they sought out and brought home to their ooloi mate or parent. In ooloi, the organ was larger and more complex. Within it, ooloi manipulated molecules of DNA more deftly than Human women manipulated bits of thread they used to sew their cloth."⁴⁰ Thus the ooloi are biologically specialized for maintaining genetic and organic materials, mediating sexual reproduction, and surveilling gene-trade mergers. Among the Oankali, ooloi are the ones who actually do the molecular engineering.

"There is evidence to show that we are recombined from powerful bacterial communities with a multibillion year history. We are part of an intricate network that comes from the original bacterial takeover of the earth."⁴¹ The Oankali are even older: Jdahya's ooloi Kahguyaht tells Lilith, "There was no life at all on your Earth when our ancestors left our original homeworld."⁴² Long before encountering Lilith's remnant of surviving humanity, before "taking over" the damaged but not entirely ruined Earth and conceiving our potential suitability for a new round of refreshing genetic recombination, the Oankali have been self-evolving for eons into a multiply metamorphic species. They will purposely engineer the corporeal transformations of their own offspring according to chance encounters with other suitable trade species and their subsequent choice or not to pursue a trade during which unforeseen complications are always possible. Thus the narration gives us to believe there are other Oankali ships trolling the galaxy as well, each with a different set of evolutionary divisions, brought about by particular gene trades that induce permanent metamorphic transformations.

The term *division* occurs a number of times in *Microcosmos*, in two senses. One is reproductive. As we have mentioned, division is the way bacterial cells reproduce: "About once in a million divisions, an offspring appears which is not identical to its parent."⁴³ The other sense is evolutionary—the formal physiological division between a prior living "kingdom" and another that evolves out of it. Margulis and Sagan assert, "so significant are bacteria and their evolution that the fundamental division in forms of life on earth is not that between plants and animals, as is commonly assumed, but between prokaryotes—organisms composed of cells with no nucleus, that is, bacteria—and eukaryotes—all the other life forms. . . . The division between bacteria and the new cells is, in fact, the most dramatic in all biology."⁴⁴ For the Oankali, as Jdahya informs Lilith, "Memory of a division is passed on biologically. I remember every one that has taken place in my family since we left the homeworld."⁴⁵

Margulis and Sagan comment how modern, technoscientific humanity's first steps in genetic engineering are a return to and appropriation of capacities bacteria have possessed for several billion years: "Our ability to make new kinds of life can be seen as the newest way in which organic memory— life's recall and activation of the past in the present—becomes more acute."[46] Their evocation of "organic memory" echoes an outmoded phase of hereditarian thought, but would appear to be intended only as a figurative way of re-marking their bacterial perspective.[47] Few ever spoke of microbes in this way before Margulis and Sagan began publishing, and without the emergence of such cultural props, humans simply don't "remember" that we go all the way back to bacteria, or what we owe them for our own capacities. But the bacteria "remember": "The bacterial world has, in effect, retained the time when the organism was not dependent on the careful packaging of DNA that occurs in nucleated organisms, but had many options open."[48]

Likely recalling the ongoing popular vogue for notions of hereditary memory, especially in science fiction, Butler may well have taken Margulis and Sagan's phrasing as a cue for a more literal appropriation. If so, at the same time Butler also inscribes Margulis and Sagan's bacterial point of view onto the bodies of the Oankali. Lilith's ooloi companion Nikanj taps its literally organic memory to recall in passing that "six divisions ago, on a white-sun water world, we lived in great shallow oceans."[49] Despite those many divisions, the Oankali recall their kinship with the beings they once were. Similarly, for us, Margulis and Sagan insist, "the microbial common denominator remains essentially unchanged. Our DNA is derived in an unbroken sequence from the same molecules in the earliest cells that formed at the edges of the first warm, shallow oceans."[50] Here with "shallow oceans" is the moment of literal textual and thematic overlap between Butler's fiction and Margulis and Sagan's exposition.

Nikanj goes directly on about his forebears' white-sun, shallow-ocean world: "We were many-bodied and spoke with body lights and color patterns among ourself and among ourselves."[51] Butler embeds in this imagery a brilliantly compact allusion to the evolutionary upshot of the life ways of the blue, green, and purple bacteria. Their propensity for genetic exchange and tinkering, their lack of immunity and sexual partiality, is precisely what allowed symbiogenetic mergers of distinct bacterial forms to evolve into the nucleated cell.⁵² Relative to bacteria, eukaryotic cells are precisely "many-bodied." But the fateful division between prokaryotes and eukaryotes was not at all divisive. Instead, it raised the stakes of bacterial recombination from genetic bits (replicons) to autopoietic unities altogether. When, after two billion years of an exclusively bacterial world, terrestrial evolution crossed this threshold, "life had moved another step, beyond the networking of free genetic transfer to the synergy of symbiosis. Separate organisms blended together, creating new wholes that were greater than the sum of their parts."⁵³ Put another way, *bacteria provide the medium out of which eukaryotic beings, all other living things, are the forms.* They are the air we breathe—their metabolism literally *makes* the air we breathe. And so, like other media, they largely disappear from view in favor of the forms they make possible.

Margulis and Sagan join other postmodern thinkers seeking to refer given forms to the contingencies of their media—to deconstruct habitual notions and factitious essences by attending to the conditions of their material possibility, their intrinsic constructedness, and the eventuality of their transformation. After Margulis, we not only see the mitochondria within all eukaryotic cells, we can also see these organelles for what they are: not pinched-off bits of our own stuff, but "the descendents of the bacteria that swam in primeval seas breathing oxygen three billion years ago."⁵⁴ Although humans host mitochondria in every cell, these organelles retain much of their own DNA within their own organellar membrane, separate from our eukaryotic nuclei. They reproduce on their own timetable and are simply handed down from mother to daughter eukaryotic cell with every mitosis. As we have established, the Oankali are an allegory of the gene-trading bacterial microcosm that supports and subsumes all other life. But Butler also draws this scenario to a head by introducing an allegory of the mitochondrion, as that organelle has become emblematic of the symbiogenetic origin of eukaryotes in general. That level of meaning first emerges in this early conversation between Jdahya and Lilith:

> "We acquire new life—seek it, investigate it, manipulate it, sort it, use it. We carry the drive to do this in a minuscule cell within a cell—a tiny organelle within every cell of our bodies. Do you understand me?"
>
> "I understand your words. Your meaning, though . . . it's as alien to me as you are."

. . . He paused. "One of the meanings of Oankali is gene trader. Another is that organelle—the essence of ourselves, the origin of ourselves."[55]

Within the narrative, the Oankali organelle is an emblem of the gene trade altogether. On its contribution to human cells pivots the xenogenetic symbiogenesis that will lead to the posthuman evolution of Lilith's brood by incorporation into the Oankali superorganism.[56] In their chapter on the emergence of the eukaryotic merger, Margulis and Sagan underscore "the wild success of the new cells . . . intertwined communities of cells-within-cells."[57] And just as the individual organisms assembled out of eukaryotic cells "are composites of species, they are also the working parts of larger superorganisms, the largest of which is the planetary patina. An organelle inside an amoeba within the intestinal tract of a mammal in the forest on this planet lives in a world within many worlds. Each provides its own frame of reference and its own reality."[58]

Formal recursion and bottomless embeddedness are the ways of all worldly systems. The merger of bacterial cells into the eukaryotic cell amounts to a metamorphosis of life forms through the feeding of cellular form back into itself. This is the neocybernetic story of microevolution from the Archean to the Proterozoic eons: a billion or two years ago bacterial life, already a billion or two years old, reentered its membrane-bounded form back into itself, like a story worth retelling beneficially placed into another more encompassing story. The mitochondrion, once a free-floating purple bacterium, became a living system within a living system. Once there are organelles within cells—nuclei, plastids, basal bodies, mitochondria, and so forth, each with its own membrane within a larger cell with its own membrane—new levels of cellular observation emerge to construct new levels of organic complexity.

We noted in chapter 4 that stories within stories suggest the embedding of worlds within worlds. Unlike many other tales of metamorphosis from *The Golden Ass* to "The Menelaiad," *The Fly*, or *The Cyberiad*, however, the Xenogenesis texts are not themselves framed or embedded; that is, they do not significantly display narrative embedding. Rather, Butler's diegesis—the storyworld itself—has been biologized: instead of stories within stories, one has organelles within cells, cells within organisms, and bodily transformations welling out of symbiogenetic embeddings. The message of the extraterrestrial beings in the Xenogenesis trilogy is in large degree the message the bacteria in *Microcosmos* have for humanity at large: "The reality and recurrence of symbiosis in evolution suggests that we are still in an invasive, 'parasitic' stage and we must slow down, share, and reunite ourselves with other beings if we are to achieve evolutionary longevity."[59]

3. ALIEN MEDIATORS

Placing the Xenogenesis trilogy into the frame of Margulis and Sagan's *Microcosmos* returns it to a fertile seminal context. The systems motifs that the Xenogenesis narrative derives from its framing by Margulis and Sagan's neocybernetic biological discourse extend to the ways that its metamorphic events are bound up with media matters. As we have argued throughout this study, posthuman metamorphs couple the media systems that enact them to the social systems communicating them. The medium—whether the words of a text, the code of a program, a narrative frame, or a bodily frame—transforms the forms it brings forth. In a more traditional idiom, literary metamorphs are allegories of writing, and narratives of metamorphosis are allegories of narrative communication. Considerations of this sort begin to coalesce in the Xenogenesis trilogy in the contrasts drawn between the modes of memory available to Oankali and humanity for ongoing social processing. "Nikanj had an eidetic memory. Perhaps all Oankali did," Lilith remarks to herself in figural narration; "anything Nikanj saw or heard, it remembered, whether it understood it or not," the authorial narrator continues: "She became ashamed of her own plodding slowness and haphazard memory."[60] Early on in her training Lilith requests and is denied writing materials. Jdahya retorts: "'Nikanj can help you remember without writing.'"[61]

Measured against the Oankali, human status is defined precisely by its defects of memory—its selective obliviousness to its environment and its history. "Remembering without writing" becomes the first threshold of the posthuman. Lilith's handlers want to strengthen her against her coming trials mediating between them and the other soon-to-be-awakened humans designated for the initiation of the gene trade. Nikanj explains: "'I must make small changes—a few small changes. I must help you reach your memories as you need them . . . a tiny alteration in your brain chemistry'"; Lilith exclaims: "'I don't want to be changed!'"[62] But she is already emotionally bonded with the adolescent ooloi. Relenting, she submits herself to its hypodermic prick, the alien vampire bite of Nikanj's sensory tentacle: "On the back of her neck, she felt the promised touch, a harder pressure, then the puncture."[63] When she awakes later and initiates conversation, it takes her a while to realize that she is speaking Oankali effortlessly. Soon, "she remembered every day that she had been awake."[64] So here is the initial stage of Lilith's posthuman metamorphosis—a vampiric transformation in which alterations remain mostly undisplayed in her bodily appearance, but which nonetheless leave a mark: she is now chemically as well as affectively bonded for life to Nikanj.

With this biosocial reassurance accomplished, Nikanj finds itself beginning *its* final metamorphosis, during which it will acquire the second pair of sensory arms of an adult ooloi. As it enters a prolonged state of semi-sleep, its family gathers—its parents, Jdahya and Tediin and their ooloi Kahguy-aht, and its mates, Dichaan and Ahajas. "The five conscious Oankali came together, touching and entangling head and body tentacles."[65] And now Lilith witnesses the counterpart to Oankali eidetic memory: just as they remember without writing, by reverting to the standard Oankali mode of neural interconnection they can communicate without speaking. Passing "messages from one to the other almost at the speed of thought" renders an image of social intimacy and solidarity that puts the fallibility of human communication in the shade: "Controlled multisensory stimulation. Lilith suspected it was the closest thing to telepathy she would ever see practiced."[66] In Butler's earlier series, the telepathic Patternists were supernaturally embodied communications devices; the Pattern itself was an unforeseen, emergent reorganization psychically networking Doro's progeny. But through the filaments of their sensory organs the Oankali and the constructs are literally wired. Oankali physiology maintains the material infrastructure of these extraterrestrials and posthumans. Their communication systems are not mystical but bacterial.

Compare that to Margulis and Sagan on genetic exchange in the microcosm: "Although exchange is easiest between bacteria that are metabolically similar, any strain can potentially receive genes from any other through a succession of intermediaries. This allows genetic information to be distributed in the microcosm with an ease and a speed approaching that of modern telecommunications."[67] In this statement Margulis and Sagan put an intriguing neocybernetic spin on their biological discourse, entertaining an informatic analogy that Butler will exploit throughout the Xenogenesis narrative. The allegory of the Oankali links bacterial metamorphosis to media technologies. But with their superhuman communications abilities grounded in an infrastructure of biological and not technological organs, the Oankali actually surpass the cyborg tropes with which Butler's exegetes usually approach them. Neither organomorphic machines nor mechano-morphic organisms, they are neocybernetic assemblages that cross over that distinction by exploiting the primeval genetic nanotechnology of living systems.

Still, in their memory and communication functions, media theorist Friedrich Kittler might read the Oankali as alien-ated media devices person-ifying the posthuman destination of the digital convergence of previously separate data streams. Even in the slumber of metamorphosis, the Oankali are always on. Regarding the seemingly unconscious metamorphosing Ni-kanj, Kahguyaht informs Lilith, "'it will remember all that happens around it. . . . It still perceives in all the ways it would if it were awake. But it cannot respond now. It is not aware now. It is . . . recording.'"[68] Seen through first-order cybernetic analogues, the ooloi in particular, designers of con-struct bodies, are self-reproducing organic computers with nervous systems that naturally undergo metamorphic upgrades. In *Imago*, the construct ooloi Jodahs remarks that after its metamorphosis, "I would soon be able to give and receive more complex multisensory illusions, and handle them much faster."[69] Genetic memories accumulated over generations are downloaded to the next: "The flood of information was incomprehensible to me at first. I received it and stored it with only a few bits of it catching my attention."[70]

In chapter 1, quoting narrative theorist F. K. Stanzel's remark about how the figural narrative situation (free indirect discourse) produces "the illusion of immediacy . . . superimposed over mediacy," we noted the tendency of media to move in and out of blind spots. The media technology of the ooloi includes a neurological analogue to this device of narrative mediation, which comes into play in their role as sexual mediators. Ooloi never mate in the sense of contributing parental genes, but all mating goes through them. Whether as an Oankali threesome or a trade fivesome, each repro-ductive cluster links itself into direct neural communication by way of the filaments their ooloi extrude to all concerned from sensory tentacles and arms. Sexed mates do not physically touch each other during sex; rather, they embrace through their ooloi mediator. At least for human participants, their ooloi transmits them multisensory illusions of direct physical contact with each other, or of virtual sexual gymnastics: "'Males in particular need to feel that they're moving,'" Nikanj tells Jodahs in *Imago*; "'You'll enjoy them more if you give them the illusion they're climbing all over you.'"[71]

Further complicating this scenario is the Oankali's unwillingness to allow straightforward human sexual reproduction to continue in their absence. According to Jodahs, although "Ooloi had examined every Human, correcting defects, slowing aging, strengthening resistance to disease," with crucial exceptions that create important complications in *Imago*, "Resisters had been altered so that they could not have children without Oankali mates."[72] Yet even those who do enter the trade are rendered unable to embrace their human mates physically without inducing a visceral aversion that effectively shortcircuits any attempt at copulation. "Only through me," Nikanj insists when Lilith reaches for her human mate Joseph.[73] Not surprisingly, when first introduced to these Oankali sexual niceties, the salvaged humans are not pleased with the arrangements. But those who accept sexual relations and their reproductive consequences on Oankali terms are rewarded with mindblowing virtual orgasms, which their ooloi "narrate" for them through a kind of figural displacement of their fictive production. Lilith "never knew whether she was receiving Nikanj's approximation of Joseph, a true transmission of what Joseph was feeling, some combination of truth and approximation, or just a pleasant fiction. . . . Nikanj focused on the intensity of their attraction, their union. It left Lilith no other sensation. It seemed, itself, to vanish."[74] In *Imago*, the construct ooloi Jodahs becomes both narrative and sexual mediator: the text as a whole is rendered through its character-bound narration. Recounting a crucial period of cross-species sexual discovery, Jodahs gives its narratee an intimate account of its technique:

> I took their hands, rested each of them on one of my thighs so that I would not have to maintain a grip. I linked into their nervous systems and brought them together as though they were touching one another. It was not illusion. They were in contact through me. Then I gave them a bit of illusion. I "vanished" for them. For a moment, they were together, holding one another. There was no one between them.[75]

4. CONSTRUCT METAMORPHOSIS

Although when I asked Butler about this during an interview in 2001, she did not confirm that she intended it, there is a pun embedded in *Oankali*—

onco-, the combining form meaning "tumor" or "mass," a prefix now signifying an association with cancer. The Oankali are onco-people—cancer-creatures. Metamorphosis in the Xenogenesis trilogy is strongly inflected by cancer; or more precisely, the metamorphosis theme here plays out in a posthuman recuperation of what for us are still the mostly catastrophic outcomes of cancer's bodily transformations. Lilith is told early on that she has a "talent for cancer,"[76] that during her long sleep, ooloi extracted her predisposition for deadly mutations while inserting "correcting genes" into her cells.[77] Part of what attracts the Oankali to the gene trade with humans is that they anticipate xenogenic benefits from Lilith's gift of cancerous cells. Jdahya spells them out: "'Regeneration of lost limbs. Controlled malleability. Future Oankali may be much less frightening to potential trade partners if they're able to reshape themselves and look more like the partners before the trade.'"[78]

For all their interstellar prowess, these Oankali do not possess such capacities when they encounter human beings. They do reshape themselves to promote the success of a gene trade, but it takes a generation: retooled individuals must be genetically engineered, then raised and nurtured through their natural, maturational metamorphoses. To tell from the current situation, when a gene trade occurs, the Oankali population involved separates into three subgroups: one stands apart altogether, in case of combinatorial catastrophe; while the construct generations develop, a second remains within the sealed Oankali environment of their organic spaceships, while a third sets out, after due preparation, to cohabit the trade species' homeworld. In Lilith's immediate experience, the first group is the Akjai, the second the Toaht, the third the Dinso. All of Lilith's intimate Oankali relations are Dinso.

One generation into the trade, one of Lilith's children—the human-born construct male Akin—travels to the ship Chkahichdahk, where he meets Kohj the Akjai:

> Akin stepped toward what his sense of smell told him was an ooloi. His sight told him it was large and caterpillarlike, covered with smooth plates. . . .
>
> It had mouth parts vaguely like those of some terrestrial insects. Even if it had possessed ears and vocal cords, it could not have formed anything close to Human or Oankali speech.
>
> Yet it was as Oankali as Dichaan or Nikanj. It was as Oankali as any intelligent being constructed by an ooloi to incorporate the Oankali organelle within its cells. As Oankali as Akin himself.

It was what the Oankali had been, one trade before they found Earth, one trade before they used their long memories and their vast store of genetic material to construct speaking, hearing, bipedal children. Children they hoped would seem more acceptable to Human tastes. The spoken language, an ancient revival, had been built in genetically. The first Human captives awakened had been used to stimulate the first bipedal children to talk—to "remember" how to talk.[79]

Here is another overlapping of the memory and mediation themes. As far as metamorphosis goes, the promise of the construct offspring of Lilith's brood will be precisely "controlled malleability," organic beings that can produce bodily self-transformations at will and at once. While, in terms of literary mythopoesis, this will be a kind of reversion to the supernatural powers of the legendary shape-shifters behind *Wild Seed*'s Anyanwu, in Xenogenesis the capacities at stake have to be earned through the bio-existential rigors—the psychic and social turmoil of the gene trade. But even before the trade generations begin at the end of *Dawn*, metamorphosis as the "regeneration of lost limbs" does occur when Lilith's ooloi mate Nikanj uses its cache of her cancer cells to survive a blow from a resister's machete that would have otherwise ended its procreative career. Instead, in another form of "awakening," Nikanj's symbiotic uploading of her "talent for cancer" into its repertoire of genetic manipulations lets it heal back its severed sensory arm, and with it, its adult ooloi functions:

> "Your body knows how to cause some of its cells to revert to an embryonic stage. It can awaken genes that most humans never use after birth. We have comparable genes that go dormant after metamorphosis. Your body showed mine how to awaken them, how to stimulate growth of cells that would not normally regenerate. The lesson was complex and painful, but very much worth learning."
>
> "You mean . . ." She frowned. "You mean my family problem with cancer, don't you?"
>
> "It isn't a problem anymore," Nikanj said, smoothing its body tentacles. "It's a gift. It has given me my life back."[80]

Although symbiogenetic merger is the encompassing frame of the entire narrative, precise matters of construct metamorphosis gestate for much of the Xenogenesis trilogy. As we have seen in *Dawn*, the Oankali give Lilith extra strengthening, but about this Nikanj asserts: " 'I haven't added or subtracted anything, but I have brought out latent ability. . . . The changes I've made are not hereditary. . . . Body cells only. Not reproductive cells.' "[81]

Construct metamorphosis begins in earnest only with the children of trade families. Oankali mating units are typically an endogamous sexual pair—a brother and sister, or at most, male and female cousins—mediated by an exogamous ooloi. Since Oankali are born *eka*, it is crucial for their individual maturation that they effectively bond to their sibling partner and achieve the psychological compatibility that ensures that their individual metamorphoses yield the requisite heterosexual distinctions. Construct offspring, possessing two human and two Oankali parents, inherit even more complex versions of these Oankali psychosexual dynamics. Episodes structured around the multiple complications generated by these complexities and other impediments to the successful metamorphoses necessary to a viable trade take up large portions of *Adulthood Rites* and *Imago*.

Adulthood Rites, set about thirty years after the time of Lilith's awakening, centers on Lilith's construct child Akin. We have noted that prior to his metamorphosis into adulthood, Akin looks human. His Oankali genetic heritage is anatomically expressed only in his tongue, which possesses sensory filaments. Like his tongue, his posthuman status remains mostly on the inside, and centers on the upgrade to his cognitive functions, such as memory: for instance, "he remembered much of his stay in the womb."[82] The emphasis in this volume is not on Akin's posthuman physique but rather the dilemmas of his hereditary differences. Butler strongly develops Akin's position as a construct interspecies child for an allegory of adolescence in an interracial family. The tensions that develop between trade families and resister communities play on racial and ethnic tensions in a polarized society reminiscent of the American South during reconstruction and its aftermath in modern ghetto culture. Akin conceives a quest for social reconciliation, and his human appearance benefits his efforts to mediate between humans and Oankali. But the sociobiological deck is stacked against him. The Oankali organelle produces a situation of genetic indeterminism, injecting metamorphic uncertainty into the living systems it parasites: "People never know what they'll be like after metamorphosis."[83] Akin's efforts to save his adopted human community from self-destruction are partially stymied when he lapses into a prolonged slumber in a resister household: "Akin's metamorphosis dragged on. He was silent and motionless for months as his body reshaped itself inside and out."[84] The arc of Akin's narrative goes from Lilith's womb to this second fetal period—his rebirth from the new gestation of his metamorphosis as an adult with an entirely Oankali body—which brings *Adulthood Rites* to an end.

Imago opens with a nine-chapter section titled "Metamorphosis." As the entire narrative winds to its climax a century into the gene trade, a prolonged and detailed depiction of construct metamorphosis fully breaks forth as the central event of the fabula. As with Akin, the trade has led to normalizing (biologizing) of metamorphic episodes on the model of adolescent changes of sexual morphology. But with the further generations of construct children from enhanced and life-extended trade mothers like Lilith, a new metamorphosis of metamorphosis is already emerging. Unlike Akin, the new beings to be perfected by the trade are to be fully functioning, deliberate (not unpredictable) metamorphs. Nikanj explains to Jodahs: "'Eventually you will awaken dormant abilities in males and females. . . . You'll be able to change yourself. What we can do from one generation to the next—changing our form, reverting to earlier forms or combinations of forms—you'll be able to do within yourself. . . . That's what we intended.'"[85]

Nevertheless, the onset of these further transformations *is* accidental in that, according to the plans Nikanj and the wider Oankali community had laid down for the trade, they occur well before they were intended.[86] In particular, as Jodahs is about to enter what should have been its normal metamorphosis resulting in the adult male everyone expects, paired to the adult female its Oankali-born *eka* sibling Aaor is expected to become, it displays "the wrong signs."[87] "I was taking on the sex of the parent I had felt most drawn to,"[88] and the source of the parental imago for Jodahs turns out to be the ooloi Nikanj. The problem is that, as the genetic engineers for the Oankali system altogether, ooloi are supposed to maintain strict control over the genetic experiments entailed by the trade, and every construct child is an ongoing experiment. Even Nikanj cannot entirely instruct Jodahs how to control the self-metamorphic powers it will spontaneously develop as a construct ooloi in maturational limbo before a second metamorphosis, after which it will surpass Nikanj's own repertoire: "Certainly nobody was ready for a Human-born construct ooloi. Could there be a more potentially deadly being? . . . A flawed natural genetic engineer—one who could distort or destroy with a touch."[89] "Its body bent in an attitude of deep shame," Nikanj confesses, "'I made a mistake.'"[90]

In the last chapter we noted how in the Western tradition of metamorphic tales, some "mistake" often triggers the transformation. A classic comedic example is Lucius in Apuleius's *The Golden Ass*, whose accomplice steals the wrong box of magical ointment, turning him not into a bird but into a jackass. Or one thinks of the play of "misprision" in *A Midsummer Night's Dream* by which Bottom ends up with an ass's head in the bed of the Fairy Queen. Additionally, the metamorphic mistake is often marked by shame: Odysseus' men in the Circe episode, Lucius in his more reflective moments, Keats' Lamia, or Kafka's Gregor, are all rendered by their narrations as registers of embarrassment and shame. I pause on these thematic analogues to underscore how the specific metamorphic motifs in *Imago* confirm the Xenogenesis trilogy's powerful connection to classical literary stories as well as folk tales of metamorphosis. In these traditions, the shame of the "metamorphic mistake" also gravitates or often alludes to imageries of incest and related monstrosities of prohibited sexual contact. The sexual and reproductive mores of the Oankali, with their admixture of endogamy and exogamy, finesse but also refract this issue. "Beyond their insertion of the Oankali organelle," Jodahs relates, the ooloi "made no genetic contribution to their children. They left their birth families and mated with strangers so that they would not be confronted with too much familiarity. . . . Male and female siblings could mate safely as long as their ooloi came from a totally different kin group."[91]

With Jodahs' incipient ooloi metamorphosis already rewiring its body chemistry, it encounters a new set of biological rather than cultural prohibitions—visceral aversions within the trade family to touching or being touched by mature or metamorphosing other-sex members except for the parental ooloi. Now, even Lilith cannot embrace her metamorphosing child, nor Jodahs she: "I had never before been unable to touch her. Now I discovered a little of the Human hunger to touch where I could not."[92] Taking on the ooloi form, in addition to shame, mistakenness, and quasi-incestuousness, Jodahs becomes associated with a related cluster of perennial metamorphic tropes; that is, it assumes a sociocultural position for which the metamorphic condition has long been a ready metaphor—social alienation and exile. Jodahs has grown up in a trade village on a piece of Earth demarcated by its implantation with the seed of an Oankali ship, the "larval version of the ship" known as Lo.[93] Past its first metamorphosis but yet unable to control the cancerous effects of its touch, Jodahs recounts how, "every day, at least, Nikanj had to correct some harm I had done to Lo. . . . Beneath me, it turned yellow. It developed swellings. Rough, diseased patches"; but even without those complications, "As an ooloi, I would have to leave it when I mated."[94] Indeed, as an exogamous extraterrestrial alien in exile, Jodahs is multiply primed for the metamorphosis of its metamorphosis.

But the full scenario of Butler's Xenogenesis narrative does not make posthuman metamorphosis any sort of seamless or superhuman proposition. The trilogy earns its neocybernetic integrity by holding the viability of the gene trade in extended suspense as these symbiogenetic constructs get their systems references aligned by negotiating a panoply of environmental, bio-logical, and social contingencies. Early in the second section of *Imago*, "Exile," on a sortie into the jungle beyond Lo, with the entire family pro-viding a buffer against scattered human desperados and resister outposts, Jodahs is in despair over its constant state of random metamorphosis: "My fingers and toes became webbed on the third day, and I didn't bother to correct them."[95] Still, several encounters with variously ailing humans show that when the occasion demands, it can concentrate its scattered metamor-phic abilities upon healing others and transforming itself into forms they find appealing: "'With a potential mate,'" Nikanj observes, "'your control is flawless.'"[96]

Yet some dire dilemmas remain. For one, Jodahs' veering away from con-struct-male identification into ooloi form has inadvertently left its paired sibling, Aaor, developmentally stranded—deprived of the intimate bond it needs for its own normal maturation. For another, only fertile mates can provide Jodahs with the conditions for establishing permanent stability as an ooloi, but no potential Oankali mates will take a chance on its unproven viability, and no potential human mates are, or at least should be, fertile. The Oankali meant to deprive all humans of mutual reproductive capacity, and despite being allowed their defiance of trade relations once resettled on Earth, the resisters can have produced no offspring of their own. Then, to compound matters, not only does Aaor go into its genetically scheduled metamorphosis, but, to compensate for the loss of Jodahs as a paired sibling, it too is turning into an ooloi and will face the same bleak prospects for a resolution of its metamorphic maturation.

The German term for metamorphosis is *Verwandlung*, or "mis-wander-ing." At a seeming impasse, with Aaor's own first metamorphosis consum-ing the exiled family's attention, Jodahs takes off: "I wandered for three days, my body green, scaly, and strange. . . . I came out of the forest, looking like nothing anyone on Earth would recognize."[97] Upon returning to a semblance of bodily normality it explains to its *eka* siblings: "'My body wanders. Even when I come home, it wants to go on wandering. . . . It's easier to do as water does: allow myself to be contained, and take on the shape of my containers.'"[98] But on its random rovings through the no-man's-land between Lo and the resister towns it eventually encounters two disfigured human wanderers lost in the forest. Tomás and Jesusa are sibling renegades from a resister village, Tomás in suicidal despair over an advanced case of neurofibromatosis (a deforming genetic disease), his less-deformed sister Jesusa in pursuit. Their illnesses fire its ooloi responses. Jodahs relates, "my body at this time was covered with fingernail-sized, overlapping scales,"[99] but it has time to transform its upper body enough not to terrify them altogether.

Capturing Tomás, it earns their acquiescence by quickly restoring mobility to his paralyzed neck. During this first encounter, Jodahs misreads their youthfulness as ooloi-induced longevity, and also misinterprets their deformities as deriving from an oversight on the part of whatever ooloi did their genetic checkup before releasing them from the trade. The next day it lets them escape, then tracks them and eavesdrops to solve the mystery of their strangeness: not only were they genuinely young, *"they were fertile. . . . I could mate with them!"*[100] Jodahs' good fortune does in fact rest on an Oankali ooloi's prior mistake, one that resolves the latter mistake by which Nikanj neglected to prevent Jodahs' choice of it as its imago. These siblings are not, as it had been raised to assume, a couple of life-prolonged but genetically mismanaged humans from the original Awakening. Rather, they are hereditarily challenged, Earth-born children of incestuous human parents—initially, as it turns out, a mother and son breeding pair—not effectively sterilized by their original ooloi. As Jodahs will soon realize: "'Your people have been breeding brother to sister and parent to child for generations.'"[101] That desperate incest was a lucky lapse, however, because only the availability of biologically youthful and fertile human mates brought about by this "mistaken" breakdown of the Oankali prohibition on non-trade human sex repairs the dilemma of the construct ooloi to enable a viable completion of the gene trade.

When Jodahs rejoins the resister couple, it displays the self-metamorphic malleability over which, once it is properly mated, it will gain full control: "'It can't be you!' he whispered when I let him speak. He remembered a scaly Jodahs, like a humanoid reptile. But I could not stay within range of their scent for four days and go on looking that way. Now I was brown-skinned and black-haired and I thought it was likely that I looked the way Tomás would when I healed him."[102] Tomás is ready to accept Jodahs' attentions, and shortly, through bodily healing and sexual touch, it breaks down Jesusa's last resistance. Promptly it finds itself entering its second metamorphosis. Tomás and Jesusa agree to raft it back to its family encampment before returning to their village, while Jodahs omits to tell them that they are now chemically bonded to it for life. It realizes it has repeated a prior act of omission, marking the inevitable imbalance of the trade partners: "Was that how Nikanj had done it a century before? Lilith had been with it when its second metamorphosis began. Had it been tempted to say, 'If you stay with me now, you'll never leave?'"[103] As Nikanj had, Jodahs says nothing. It seizes its chance to establish the human component of a mating unit that can consummate the reproductive independence of the construct posthumans.

Once Jodahs is reunited with Lilith, Nikanj, and the rest, the scene of metamorphic wandering shifts to Aaor, who has come out of its first metamorphosis in the same fix that Jodahs has just resolved. Starved for touch, Aaor slips into morphic dissolution, allowing the narrative one more tour de force in the meta-metamorphic mode. Even more than Jodahs, Aaor puts a posthuman spin on the traditional metamorphic romance, in which the human body changed into nonhuman form represents an existential exile, doubled by a territorial exile—a wandering in the wilderness—resolved when the metamorph finds or is granted means to return "home," in both senses. But, for the moment, all Aaor can do is wander ever farther away:

> It changed radically: grew fur again, lost it, developed scales, lost them, developed something very like tree bark, lost that, then changed completely, lost its limbs, and went into a tributary of our river.
>
> When it realized it could not force itself back to a Human or Oankali form, could not even becomes a creature of the land again, it swam home. . . .
>
> Hozh showed me what Aaor had become—a kind of near mollusk, something that had no bones left. Its sensory tentacles were intact, but it no longer had eyes or other Human sensory organs. Its skin, very smooth, was protected by a coating of slime.[104]

Aaor's devolving metamorphoses are a kind of biological entropy, a break-down of the increasing complexity otherwise being engineered by the directed evolution of the gene trade. These construct ooloi devolve as readily as evolve, since for the Oankali there is neither an essential "nature" nor any final term or evolutionary *telos* by which to measure or rank the sequence of their "divisions."

The Oankali are naturally neocybernetic, enjoying a constructivist *Welt-anschauung* that transcends ontological fixations of any sort. But, for this very reason, the maintenance of their social autopoiesis through creaturely tactility and neural communications is imperative. As Eric White astutely remarks on the Oankali: "Structural complexity, and the consequent possibility of further differentiation and metamorphosis, depends on their being situated in a social matrix less chaotically mutable than themselves."[105] Aaor suffers precisely from the current deficit in that social matrix. Entering its second metamorphosis without exogamous mates, neither its ooloi parent or sibling nor its entire family matrix can finally stabilize its biological death drive. Like a heroin-addicted child under parental surveillance, "again and again, it had to be brought back from drifting toward dissolution"; Nikanj complains, " 'The moment I release it, it drifts toward a less complex form.' "[106]

Aaor's organic extremity threatens to unravel a symbiogenetic assemblage: "If it had stayed unattended in the water for much longer, it would have begun to break down completely—individual cells each with its own seed of life, its own Oankali organelle . . . but Aaor as an individual would be gone."[107] As an Oankali-born construct, Aaor's condition may be a symbolic repercussion of its onco-association—its devolving metamorphosis may be the allegory of an unrecuperated cancer. Margulis and Sagan comment that during the cellular upheavals of a growing cancer, "it is as if the uneasy alliances of the symbiotic partnerships that maintain the cells disintegrate. The symbionts fall out of line, once again asserting their independent tendencies, reliving their ancient past. . . . Cancer seems more an untimely regression than a disease."[108] And once again, as Lilith had done for Nikanj at the end of *Dawn*, a captive human female must step forward to restore an ailing parasexual alien. This time the savior of the ooloi is Jesusa, who against her own engrained revulsion at leading other humans to betray the species by joining the trade, finds the fund of cross-species sympathy for Aaor's suffering necessary to enquire: " 'If Aaor had a chance to mate with a pair of Humans . . . would it survive?' "[109]

The essential claim of the sublime is that man can, in feeling and in speech, transcend the human.

— THOMAS WEISKEL

Thus is conceived the penultimate episode of *Lilith's Brood*–the expedition Jodahs and Aaor make with Jesusa and Tomás to their mountain village. Aaor gets his sexual fix, and eventually the construct duo seduces with pheromones and wins over with healing powers a population of fertile, genetically reparable, Earth-born humans sufficient to supply the Dinso, Toaht, and their constructs with enough trade mates to secure the success of this ongoing division of the Oankali. We will end, however, with two posthuman moments that frame that episode. The first is the biocultural transfer between Nikanj and Jodahs we glimpsed in section 3, a last reprise of the way this narrative crosses organic memory with echoes of media technology. Nikanj uses the Oankali term of endearment for adolescents as an affectionate diminutive inflecting the completed adult Jodahs has almost become:

> "Lelka, I have memories to give you. . . . Let me pass them to you now. I think it's time."
>
> Genetics memories.[110] Viable copies of cells that Nikanj had received from its own ooloi parent or that it had collected itself or accepted from its mates and children. It had duplicated everything it possessed and now it would pass the whole inheritance on to me. It was time. I was a mated adult. . . .
>
> Then the world around me seemed to flare brilliant white. I could no longer see beyond myself. All my senses turned inward as Nikanj used both sensory hands to inject a rush of individual cells, each one a plan by which a whole living entity could be constructed. The cells went straight into my newly mature yashi. The organ seemed to gulp and suckle the way I had once at my mother's breast.
>
> There was immense newness. Life in more varieties than I possibly could have imagined—unique units of life, most never seen on Earth. Generations of memory to be examined, memorized. . . .
>
> The flood of information was incomprehensible to me at first. I received it and stored it with only a few bits of it catching my attention.[111]

The climax of Butler's science-fictional allegory of the informatic microcosmos offers a wonderful reprise of the Romantic sublime, in particular those liminal moments when the poetic ephebe receives an overwhelming transmission of divine or worldly knowledge. In Book 3 of John Keats' *Hyperion* (1819), Apollo stands on the threshold of his celestial vocation, and the one who has memories to give him is none other than Mnemosyne, the goddess of memory:

> Mute thou remainest—mute! yet I can read
> A wondrous lesson in thy silent face:
> Knowledge enormous makes a God of me.
> Names, deeds, gray legends, dire events, rebellions,
> Majesties, sovran voices, agonies,
> Creations and destroyings, all at once
> Pour into the wide hollows of my brain,
> And deify me.[112]

Walt Whitman's "Out of the Cradle Endlessly Rocking" (1859) tosses off Keats' Hellenistic vestments. Whitman puts his autobiographical ephebe on the strand of Long Island and channels the natural sublime through the cries of a seabird:

> Demon or bird! (said the boy's soul,)
> Is it indeed toward your mate you sing? or is it really to me?
> For I, that was a child, my tongue's use sleeping, now I have heard you,
> Now in a moment I know what I am for, I awake,
> And already a thousand singers, a thousand songs, clearer, louder and more
> sorrowful than yours,
> A thousand warbling echoes have started to life within me, never to die.[113]

Jodahs' post-metamorphic Awakening to comprehensive biocultural comprehension marks its assumption of the *imago* or adult stage of a new construct species: "We represented the premature adulthood of a new species. We represented true independence—reproductive independence—for that species."[114] Because it channels through a singular receptacle both the cosmic history and knowledge of an entire species and the breakthrough to a viable transformation of that species, and because the human species finds its consummation here by absorption into a transcendental posthuman consortium, one could call this climactic moment of the narrative the evocation of an *evolutionary* sublime.

But the posthuman sublime depicted here at the climax of the Xenogenesis trilogy transcends the human in a register different than, in Thomas Weiskel's formulation, "feeling" or "speech." Those Romantic registers grounded the sublime in a breakdown of the couplings of psychic and/or social systems to their respective environments, when perception (mind or feeling) or communication (or speech) "rise above" their material media.[115] In Butler's biological or cellular framing of this occasion, the "transmission" that both dissolves the individual Jodahs into its collective living networks and confirms its viable autonomy as a capable receiver of this message is carried out (as far as one can speculate here) immanently—in the medium of organic macromolecules. Whatever metabiotic meanings arise for Jodahs in its consciousness and then through its (narrative) communications would be secondary constructions of a consummately biotic intercourse.

The master key to what Butler's narrative has accomplished, then, is the way that she channels all that macrocosmic extraterrestrial transcendentalism through the microcosm, "a rush of individual cells." Nor are we left just with a vista of Jodahs as the culture hero of construct biological futurity. The cellular sublime is transmitted not just through a singular posthuman "individual," but through its care and nurture of a basal biotic individual, a single cell: "I sorted through the vast genetic memory that Nikanj had given me. There was a single cell within that great store—a cell that could be 'awakened' from its stasis within yashi and stimulated to divide and grow into a kind of seed. . . . My seed would begin as a town and eventually leave Earth as a great ship."[116] The Xenogenesis trilogy ends with a primal scene of posthuman reproductive technology aimed at what Margulis and Sagan call the "supercosm."[117] The means by which life on Earth will eventually take to the stars is rooted firmly in the fertile humus of an Earth sculpted for billions of years by the autopoietic purposes of operationally autonomous but biospherically interconnected living cells:

> I prepared the seed to go into the ground. I gave it a thick, nutritious coating, then brought it out of my body through my right sensory hand. I planted it deep in the rich soil of the riverbank. Seconds after I had expelled it, I felt it begin the tiny positioning movements of independent life.[118]

The Neocybernetic Posthuman

To return a last time to Latour's term, narratives of bodily transformation are nonmodern—at once archaic and posthuman. The narrative metamorphs of every era are allegorical beings that index systemic complexes. Their altered bodies convey the materialities of their own mediated being and the forms of the psychic and social systems in the environments to which their media couple them. Narrative mythopoesis is a nonmodern reflection on the human as a nexus for a complex embedding of systems and environments with operational concurrence but without overriding operational unity. Despite the garish predictions of cyberdigital gurus, there will be no lingua franca or metacode into which all corporeal and systemic phenomena can ultimately be translated.[1] Visions of organic abandonment through digital convergence mystify the posthuman.

In contrast, with Luhmann the neocybernetic posthuman starts here: "There is no fundamental common ground among systems."[2] What is the case is a common relation of difference: the system–environment distinction. Autopoietic systems of whatever kind must observe their environments across operational boundaries. But the operations that carry observations out are strictly internal to each system. Luhmann continues about the operational closure of system processes in *Social Systems*: "No system unity can exist between mechanical and conscious operations, between chemical operations and those that communicate meaning. There are machines, chemical systems, living systems, conscious systems, and (social) systems that communicate via meaning; but no system unities encompass all these at once. A human being may appear to himself or to an observer as a unity, but he is not a system."[3] Posthumanism cognizes the human as one among numberless other situations of complexity—a productive disunity tasked with the quest, different for every psychic and social system, of working out a viable coordination of its systemic and environmental multiplicities.

In this study we have contemplated fictive images of systemic merger—the classical image of the human along with its perennial mythopoetic projection, the metamorph—as allegories of the reality of systems distinctions. The emergence of languages, of mythic and literary narratives, are primordial social responses to the forms of limitation and inaccessibility instituted by the necessary separation yet functional combination or structural coupling of distinct yet interpenetrated, coevolving systems. The basic gesture of mythopoetic metamorphoses is the construction of images and narratives of merger among functionally coupled but operationally closed systems:

- the merger of physical and psychic systems in the form of gods, spirits, and demons
- the merger of living and psychic systems in theology's immortal soul
- the merger of distinct psychic systems (souls) in the erotic sublime
- the merger of physical and living systems in vitalism's life force
- the merger of physical, neural, and mechanical systems in the robot
- the merger of living, psychic, and mechanical systems in the cyborg
- the merger of distinct genetic systems (species) in the metamorph

The system reference that goes unmarked in each instance is the social system, for which each mythopoetic cluster functions as a viable communicative offer, generally in a narrative medium. The neocybernetic turn on these fabulations is to understand such assemblages not as images of fusion that reduce the multiple to unity, but as constructions that achieve narrative viability as complex signs of interpenetrations that integrate while maintaining systemic differences. On the level of biotic systems, Margulis and Sagan have been eloquently arguing for over two decades about the natural metamorphoses of biological evolution: "All organisms of greater morphological complexity than bacteria, that is, nucleated or eukaryotic organisms (whether single-celled or multicellular), are also *poly*genomic. They have selves of multiple origins . . . comprised of *hetero*logous different-sourced genomic systems that each evolved from *more than one* kind of ancestor."[4] On the level of metabiotic systems, Luhmann writes: "Consciousness compensates for the operative closure of the nervous system, just as the social system compensates for the closure of consciousness."[5] According to the best scientific and social theorizing at my disposal, the discrete merger of separate systems into hybrid consortiums is the way the world works. In that case, stories about imaginary versions of them are effective narrative compensations for the contingencies of systemic closure that remain once the mergers have been accomplished.

Mythopoetic assemblages from the Sphinx to the Brundlefly to the construct offspring of humans and aliens convey images of the necessary hybridity—the multi-sidedness—of any form of identity. The neocybernetic posthuman joins with Archean evolution and premodern premonitions of bodily metamorphosis to observe that the noise and heterogeneity generated by the self-maintenance and self-reproduction of systems, starting with living systems, are the price of their ongoing existence as well as the cause of their eventual cessation or subsumption into posterior forms.

Human technologies have produced a hypercomplex environment for which humanist distinctions between the natural, the human, and the technological are increasingly nonfunctional. Cybernetics has allowed us to embed mechanisms within our bodies and to insert vast mechanical and computational systems into the world around us. But regardless of the extent to which we impose such devices upon ourselves and the planet, high-tech instrumentality in and by itself does not transcend the human. A notion of the posthuman worth the name, capable of affirming our ongoing viability as a species, begins once we coordinate our systems with the geobiological world's instrumentalities as well, as it were, (meta)symbiotically. The organic bodies and ecosystems we impose our technologies on are not beneath us but *beyond* us, even while all around us, even while sharing us with an environment as yet fit for life. Whether it wants to or not, humanity will have to post itself to the Gaian conception of its embeddedness within geobiological phenomena that are planetary and cosmic in scope. It will earn its continuation only by metamorphic integration into new evolutionary syntheses.

No system can subdue or contain the entirety of its environment. Systems are possible only within environments that entirely surpass them. Nature and technology share similar systematic contingencies regarding the boundaries that enclose the operation of systems and embed them within contexts greater than themselves. Thus the posthuman does not transcend the human as the discourse of the human has imagined transcendence. Rather, the neocybernetic posthuman transcends the vision of disconnection that has isolated the human for so long in its own conceit of uniqueness. The reconnections called for will not be fusions that dissolve autonomies but couplings preserving operational differences. The neocybernetic posthuman is the human metamorphosed by reconnection to the worldly and systemic conditions of its evolutionary possibility.

1. This is the overall thesis of my previous study of the Western literature of metamorphosis from classical to modern times, *Allegories of Writing: The Subject of Metamorphosis*.

2. See Halberstam and Livingston, ed., *Posthuman Bodies*; Badmington, ed., *Posthumanism*; Lenoir, ed., "Makeover: Writing the Body into the Posthuman Technoscape, Part One: Embracing the Posthuman" and "Part Two: Corporeal Axiomatics"; Badmington, ed., "Posthuman Conditions"; Graham, *Representations of the Post/Human: Monsters, Aliens, and Others in Popular Culture*; and Weinstone, *Avatar Bodies: A Tantra for Posthumanism*.

3. Latour, *Aramis, or the Love of Technology*, 227.

4. See Hardison, *Disappearing Through the Skylight: Culture and Technology in the Twentieth Century*.

5. Hayles, *How We Became Posthuman: Virtual Bodies in Cybernetics, Literature, and Informatics*, 3.

6. Kittler, *Gramophone, Film, Typewriter*. See Clarke, "Friedrich Kittler's Technosublime."

7. See Tabbi, ed., *Reading Matters: Narratives in the New Media Ecology*, esp. Paulson, "The Literary Canon in the Age of its Technological Obsolescence," 227–49.

8. For a standard critique of the "radical technophilia" associated with the main line of first-order cybernetics, see Bendle, "Teleportation, Cyborgs and the Posthuman Ideology." More to the point is Rossini, "Figurations of Posthumanity in Contemporary Science/Fiction: All Too Human(ist)?"

9. See Hayles, *Posthuman*; DuPuy, *The Mechanization of the Mind: On the Origins of Cognitive Science*; and von Foerster, *Understanding Systems: Conversations on Epistemology and Ethics*, 135–40.

10. Brand, "For God's Sake, Margaret: Conversation with Gregory Bateson and Margaret Mead," 33.

11. Bateson's *Steps to an Ecology of Mind* stands at the threshold of the second-order turn. On second-order cybernetics in technoscientific context, see Heylighen, "Cybernetics and Second-Order Cybernetics." For a range of approaches connecting second-order cybernetics to disciplines across the humanities and the cognitive and social sciences, see Serres, *The Parasite*; Paulson, *The Noise of Culture: Literary Texts in a World of Information*; all of Niklas Luhmann's work: his most comprehensive statement currently in English translation is *Social Systems*; Wellbery, ed., "Observation, Difference, Form: Literary Studies and Second-Order Cybernetics"; Wolfe, "Systems Theory: Maturana and Varela with Luhmann"; Rasch, *Niklas Luhmann's Modernity: The Paradoxes of Differentiation*; Tabbi, *Cognitive Fictions*; Livingston, *Between Science and Literature: An Introduction to Autopoetics*; and Moeller, *Luhmann Explained: From Souls to Systems*.

12. Von Foerster, "On Self-Organizing Systems and Their Environments." See also Brier, "The Construction of Information and Communication: A Cybersemiotic Reentry into Heinz von Foerster's Metaphysical Construction of Second-Order Cybernetics."

13. Working from Henri Atlan's adaptations of von Foerster, Serres cultivates *le parasite*, his figure for the transformative agency of informatic noise and for the observer who can add signal and noise together. In addition to *The Parasite*, see "Platonic Dialogue" and "The Origin of Language: Biology, Information Theory, and Thermodynamics," in Serres, *Hermes*, 65–83.

14. See Maturana, *The Tree of Knowledge: The Biological Roots of Human Understanding*.

15. The convergence of von Foerster and Luhmann on a problem opened by Maturana and Varela—the "composition" of multiple autopoietic systems—is particularly clear in von Foerster, "Luhmann."

16. Luhmann, "The Cognitive Program of Constructivism and a Reality That Remains Unknown," 147.

17. On trivial and nontrivial machines, see von Foerster, "Luhmann"; and Niklas Luhmann, "The Control of Intransparency." We will return to these matters in chapter 5.

18. This usage is drawn from Luhmann's text: see "I See Something You Don't See," 190. According to von Foerster, "in cybernetics you learn that paradox is not bad for you, but it is good for you, if you take the dynamics of the paradox seriously." Von Foerster, "Interview."

19. Latour, *Aramis*, 119.

20. Haraway, "A Cyborg Manifesto: Science, Technology, and Socialist-Feminism in the Late Twentieth Century."

21. Haraway, "Cyborgs and Symbionts: Living Together in the New World Order." Rosi Braidotti assesses Haraway's theoretical contributions in "Posthuman, All Too Human: Towards a New Process Ontology."

22. Instrumental in bringing a wide academic audience to Butler's science fiction, Haraway's "Manifesto" highlighted *Dawn*, the first volume of the Xenogenesis trilogy. About the first edition of Margulis and Sagan's *What is Life?* Haraway writes: "These writers cross-stitch technology, organic beings, and inorganic nature into a cobbled together, profoundly materialist and dynamic biosphere. . . . Margulis and Sagan provide an historical narrative with a future that is full of metamorphoses, but without apocalypses" ("Cyborgs," xvii).

CHAPTER 1: NARRATIVE AND SYSTEMS

1. Lyotard, *The Postmodern Condition: A Report on Knowledge*, 37. Lyotard touches on Luhmann later in this text. For context and detailed criticism, see Rasch, *Luhmann*, esp. "In Search of the Lyotard Archipelago," 84–107.

2. Luhmann, *Observations on Modernity*, ix.

3. Ibid., x.

4. Jean Baudrillard cited in Winthrop-Young, "Translators' Introduction: Friedrich Kittler and Media Discourse Analysis," xv.

5. For a reformulation of Luhmann's position by way of Baudrillard, see Bjerg, "Accelerating Luhmann: Towards a Systems Theory of Ambivalence."

6. Latour, *We Have Never Been Modern*. For a current positioning of Latour's argument, see Venn, "Modernity."

7. For more on the conceptual interplay of Latour and Luhmann, see Clarke, "Strong Constructivism: Modernity and Complexity in Science Studies and Systems Theory."

8. Habermas, *The Philosophical Discourse of Modernity: Twelve Lectures*. See Moeller, *Luhmann Explained*, 187–91.

9. Habermas, *Philosophical Discourse*, 373 [sic]. As translated, the antecedent of "it" is ambiguous because the immediate candidates are plural in construction. Presumably the antecedent is presented by the phrase "logocentric limitations," which, according to Habermas, systems theory does not surpass but instead subverts.

10. Luhmann, "The Autopoiesis of Social Systems," 172.

11. Latour, *Aramis*, 222.

12. For an informative overview of the field, see Cobley, "Narratology." Cobley presents a concise history of the genre in *Narrative*.

13. Abbott, *The Cambridge Introduction to Narrative*, 12–13.

14. Genette, *Narrative Discourse: An Essay on Method*, 25.

15. Ibid., 25.

16. Ibid., 26.

17. Ibid., 28–29.

18. Phelan, "Narrative Theory, 1966–2006: A Narrative," 283.

19. Bal, *Narratology: An Introduction to the Theory of Narrative*, 6.

20. Ibid., 5–6.

21. See Ryan, *Avatars of Story*.

22. Bal, *Narratology*, 6, 8, 78.

23. Ibid., 8, 79.

24. Abbott cites Jonathan Culler on the "'double logic' of narrative, since at one and the same time story appears both to precede and come after narrative discourse" (*Introduction to Narrative*, 18), referencing Culler, *The Pursuit of Signs: Semiotics, Literature, Deconstruction*.

25. Bal, *Narratology*, 9.

26. Luhmann, "What is Communication?" 157.

27. Ibid., 160.

28. Luhmann, "The Medium of Art," 216.

29. Ibid., 216.

30. Luhmann, "Paradox of Observing Systems," 84.

31. Ibid., 85.

32. Ibid., 84.

33. Bolter, *Remediation: Understanding New Media*, 5.

34. Ibid., 5–6.

35. Bernard Duyfhuizen terms an equivalent concept of mediacy "narrative transmission," in *Narratives of Transmission*, 5.

36. Stanzel, *A Theory of Narrative*, 5.

37. Ibid., 9.

38. Ibid.

39. Butler, *Lilith's Brood*, 7.

40. Genette, *Narrative Discourse*, 186.

41. Ibid., 192.

42. Bal, *Narratology*, 142.

43. Maturana, *Tree of Knowledge*, 26.

44. Ibid., 146.

45. Prince, "A Point of View on Point of View or Refocusing Focalization,"
46. In my own text I have adopted Bal's orthography "focalizor" (on the model of narrator) for the agent of focalization.

46. Phelan, "Why Narrators Can Be Focalizers—And Why It Matters," 51.

47. Ibid., 57.

48. Bal, *Narratology*, 16.

49. Ibid., 47.

50. Prince, "Point of View," 47.

51. Bal, *Narratology*, 161–62.

52. Ibid., 163.

53. Ibid., 193.

54. Ibid., 79.

55. Ryan, *Avatars*, 7

56. Ibid.

57. Cobley, "Narratology," 678.

58. See Herman, ed., *Narrative Theory and the Cognitive Sciences.*

59. Abbot, *Introduction to Narrative*, 17.

60. Luhmann, "Communication," 161.

61. Ibid., 160.

62. Ibid., 156.

63. Shakespeare, *A Midsummer Night's Dream*, 5.1.23–26.

64. Bal, *Narratology*, 175.

65. Ibid., 177–78.

66. Ryan, *Avatars*, 102.

67. Wells, *The War of the Worlds*, 3.

68. See Markley, *Dying Planet: Mars in Science and the Imagination.*

69. Wells, *War of the Worlds*, 138.

70. "A special case of focalization and perhaps the best justification for the distinction I am making [between narrator and focalizor] is memory. Memory is an act of 'vision' of the past but, as an act, situated in the present of the memory. . . . The 'story' the person remembers is not identical to the one she experienced." (Bal, *Narratology*, 147).

71. Wells, *War of the Worlds*, 191.

72. Arthur C. Clarke, *Childhood's End*, 6.

73. Ibid., 211–212.

74. We will return to the topic of narrative telepathy at the end of the section "Narrative Situations" in chapter 3.

75. Sagan, *Contact*, 4.

76. Ibid, 17.

77. Ibid., 56–57.

78. Ibid., 91.

79. Ibid., 334.

80. Ibid., 359.

81. Luhmann, "The World Society as a Social System," 131.

82. For another view of SETI from a communications perspective, see Peters, "Communication with Aliens," in *Speaking into the Air: A History of the Idea of Communication*, 246–57.

83. Luhmann, "World Society," 131–32.

CHAPTER 2: NONMODERN METAMORPHOSIS

1. See Serres, *Genesis*, esp. 87–91.

2. See Latour, *Laboratory Life: The Construction of Scientific Facts*; and Latour, *Science in Action: How to Follow Scientists and Engineers through Society.* The quasi-object often emerges in Latour's text in the two-sided form "quasi-objects, quasi-subjects." He introduces "these strange new hybrids . . . what, following Michel Serres (1987), I shall call quasi-objects, quasi-subjects" (*Modern*, 51). For more on this collaboration, see Serres, *Conversations on Science, Culture, and*

Time; Wesling, "Michel Serres, Bruno Latour, and the Edges of Historical Periods"; and Clarke, "Science, Theory, and Systems." On the quasi-object in feminist science studies, see Haraway, "The Promises of Monsters: A Regenerative Politics of Inappropriate/d Others"; and Squier, "From Omega to Mr. Adam: The Importance of Literature for Feminist Science Studies."

3. Wells, *The Island of Dr. Moreau*, 104. Squier reads *Moreau* along with comparable biological fantasies in "Interspecies Reproduction: Xenogenic Desire and the Feminist Implication of Hybrids." On Darwinian issues, see Krumm, "*The Island of Dr. Moreau*, or the Case of Devolution." See also Gold, "Reproducing Empire: *Moreau* and Others."

4. Latour, *Aramis*, 173.

5. Calvino, "Cybernetics and Ghosts," 5.

6. Luhmann, *Art as a Social System*, 10.

7. The neocybernetic literature on the role of paradox in the operation of meaning systems (psychic and social systems) is detailed in chapter 3.

8. While Latour's work has been taken up extensively in critical and theoretical discussion across the humanities and social sciences, he has received less notice in literary criticism. Notable exceptions are literature and science scholars Crawford, "Imaging the Human Body: Quasi Objects, Quasi Texts, and the Theater of Proof," and "Networking the (Non) Human: *Moby-Dick*, Matthew Fontaine Maury, and Bruno Latour"; and Squier, "Omega." On Latour's influence on interdisciplinary science studies in the humanities, see the discussions of *inscription* in Clarke, ed., *From Energy to Information: Representation in Science and Technology, Art, and Literature*. For more on Latour in relation to systems theory, see Clarke, "Strong Constructivism."

9. Latour relates the repressed productions of modern mediations to Donna Haraway's readings of the "material-semiotic practices and entities" of technoscience: "The [modern] Constitution explained everything, but only by leaving out what was in the middle. 'It's nothing, nothing at all,' it said of the networks, 'merely residue.' Now hybrids, monsters—what Donna Haraway calls 'cyborgs' and 'tricksters' (Haraway, 1991) whose explanation [the modern Constitution] abandons—are just about everything" (*Modern*, 47). For Haraway, the cyborg is an emblem within a practice of technoscientific figuration akin to Latour's brand of science studies, an emblem she has infused with a gender dynamic generally absent from Latour's discourse: "The implosion of the technical, organic, political, economic, oneiric, and textual that is evident in the material-semiotic practices and entities of late-twentieth-century technoscience informs my practice of figuration. Cyborg figures—such as the end-of-the-millennium seed, chip, gene, database, bomb, fetus, race, brain, and ecosystem—are the offspring of the implosions of subject and objects and of the natural and artificial" (Haraway, *Modest_Witness*, 12). See also her notice of debt to and critique of Latour in "The Promises of Monsters."

10. Latour, *Aramis*, 131

11. Latour, *Science in Action*, 132–34.

12. The defensiveness and injustice dispensed to Latour by the "science warriors" who set forth over a decade ago to sound the alarm about "postmodern anti-science" get an adequate and amusing journalistic account in Berreby, ". . . that damned elusive Bruno Latour." A more balanced critique is Kukla, *Social Constructivism and the Philosophy of Science.* Latour's *Pandora's Hope: Essays on the Reality of Science Studies* is a rebuttal to those critics; see especially chapters 7 and 8 (216–65). See also Latour's self-critique in "The Promises of Constructivism."

13. Latour, *Modern*, 3.

14. Ibid., 10–11.

15. Ibid., 12.

16. Ibid., 37.

17. Ibid.

18. "An intermediary—although recognized as necessary—simply transports, transfers, transmits energy from one of the poles of the Constitution. It is void in itself and can only be less faithful or more or less opaque. A mediator, however, is an original event and creates what it translates as well as the entities between which it plays the mediating role" (Latour, *Modern*, 77–78). Readers of Serres will also recognize here the *parasite.*

19. Latour, *Modern*, 63.

20. Ibid., 64.

21. See Clarke, *Allegories of Writing.*

22. Latour, *Modern*, 64.

23. Ibid., 81.

24. Ibid., 129.

25. On communication as a concept moving across the animal/human division, see Wolfe, "In the Shadow of Wittgenstein's Lion: Language, Ethics, and the Question of the Animal."

26. *Latour, Modern*, 129.

27. Ibid.

28. "No one has ever heard of a collective that did not mobilize heaven and earth in its composition, along with bodies and souls, property and law, gods and ancestors, powers and beliefs, beasts and fictional beings. . . . Such is the ancient anthropological matrix, the one we have never abandoned. But this common matrix defines only the point of departure of comparative anthropology" (Latour, *Modern*, 107).

29. Latour, *Modern*, 136.

30. Ibid., 137.

31. Ibid., 138.

32. Charles Darwin, *The Origin of Species*, 172.

33. Wells, *The Island of Dr. Moreau*, 53.

34. Ibid., 24, 29.
35. Ibid., 62.
36. Ibid., 61.
37. Ibid., 43.
38. Ibid., 59.
39. Ibid., 57.
40. Ibid., 58.
41. Latour, *Aramis*, 118.
42. Ibid., 118, 119.
43. Ibid., 119.
44. Ibid., 173.
45. Ibid., 227.
46. Latour, *Modern*, 115.

CHAPTER 3: SYSTEM AND FORM

1. Serres, *Conversations*, 60.
2. Luhmann, *Social Systems*, 60.
3. Niklas Luhmann's most extended treatments of literary matters currently in English translation are found in *Love as Passion: The Codification of Intimacy*, but see also Luhmann, *The Reality of the Mass Media*, esp. chapter 5, "Entertainment," 51–62. Systems–theoretical approaches to literature broadly considered are presented or discussed in Schmidt, "Literary Systems as Self-Organizing Systems"; Udwin, "Autopoiesis and Poetics"; Halpern, "The Lyric in the Field of Information: Autopoiesis and History in Donne's *Songs and Sonnets*"; Roberts, "Self-Reference in Literature"; Hayles, "Making the Cut: The Interplay of Narrative and System, or What Systems Theory Can't See"; Nünning, "On the Perspective Structure of Narrative Texts: Steps Toward a Constructivist Narratology"; Reinfandt, ed., *Systems Theory and Literature*; McMurry, *Environmental Renaissance: Emerson, Thoreau, and the American System of Nature*; and Economides, "'Mont Blanc' and the Sublimity of Materiality." See also de Berg, "Luhmann's Systems Theory and Its Applications in Literary Studies: A Bibliography."
4. Ashby, *An Introduction to Cybernetics*, 2.
5. See Levinson, "What is New Formalism?"
6. Luhmann, *Art*, 38.
7. See figure 15 in Hughes, *Vicious Circles and Infinity: An Anthology of Paradoxes*.
8. We will return to this literature in chapter 4 in reference to narrative frames and embedding. Many considerations of literary paradox (as well as narrative embedding) draw from the work of Jorge Luis Borges, who investigates paradox both in fiction, such as "The Circular Ruins," and nonfiction, such as

"Avatars of The Tortoise." On Borges in this context, see especially Merrell, *Unthinking Thinking: Jorge Luis Borges, Mathematics, and the New Physics*, esp. chapter 2, "A Predilection for Paradox," 32–52.

9. See von Foerster's remarks on the history of paradox in his Stanford Lectures on CD-ROM, "A Constructivist Epistemology" (1983), in *Heinz von Foerster 90*, ed. Grössing.

10. Luhmann, "Paradox," 80.

11. Ibid., 80–81.

12. For a post-Spencer-Brownian logic reconstructed for recursivity, see Hellerstein, *Diamond: A Paradox Logic*.

13. *Laws of Form* develops a "non-numerical mathematics" that Spencer-Brown provides with a philosophical gloss. My presentation of it here, as in Luhmann's school generally, is non-mathematical, and in my own case, occasionally impressionistic. John Mingers rehearses the laws of form in relation to biological and computational autopoiesis in "Mathematics and Models for Autopoiesis," chapter 4 of *Self-Producing Systems: Implications and Applications of Autopoiesis*. See also the remarks on Spencer-Brown and paradox in Wilden, *System and Structure: Essays in Communication and Exchange*; and the adaptation of the laws of form to semiotics in Merrell, *Semiotic Foundations: Steps toward an Epistemology of Written Texts*. The wider appropriation of Spencer-Brown by Luhmann and his school is well represented in *Problems of Form*, ed. Baecker. Schiltz's "Space is the Place: The *Laws of Form* and Social Systems" is both systems–theoretical and mathematically informed.

14. "Very unusual and contentious work": Mingers, *Self-Producing Systems*, 49. Von Foerster assisted the initial recognition of *Laws of Form* with an enthusiastic review, along with a sampling of Spencer-Brown's philosophical commentaries on his formal system. See von Foerster, "*Laws of Form*."

15. Louis H. Kauffman, "Self-Reference and Recursive Forms," 53.

16. Luhmann, "Cognitive Program," in *Theories*, ed. Rasch, 132.

17. Spencer-Brown, *Laws of Form*, 1.

18. Ibid., 2.

19. Recent commentary has concentrated on the way this image encodes Western gender ideology: here observation shows forth as a male prerogative, while the female is fully exposed as the object of that controlling and defining gaze. See Haraway, *Modest_Witness*, 180ff., and Bolter, *Remediation*, 78–79. Systems–theoretical form analysis underwrites the paradoxical and exclusionary powers of sex and gender distinctions without itself privileging those distinctions. For a Luhmannian approach to gender issues, see Cornell, "Enabling Paradoxes: Gender Difference and Systems Theory."

20. Von Foerster, "*Laws of Form*."

21. "What we should never forget is that one of the central intentions of the study of autopoiesis and organizational closure is to describe a system with

no inputs or outputs (which embody their control or constraints) and to empha-
size their autonomous constitutions," in Varela, *Principles of Biological Autonomy*,
56.

22. In *Social Systems*, Luhmann stresses the "temporalized complexity" of
autopoietic meaning systems: "The theory of temporalization's most impressive
consequence is that a new *interdependence of the disintegration and reproduction*
of elements results. Systems with temporalized complexity *depend on constant
disintegration*. Continuous disintegration creates, as it were, a place and a need
for successive elements" (48).

23. Roberts, "Self-Reference in Literature," 40.

24. Shakespeare, *A Midsummer Night's Dream*, 1.1.234–35.

25. Luhmann, *Art*, 48.

26. Shakespeare, *Midsummer*, 2.1.123–27.

27. Ibid., 2.1.163–68. On the allegory of Elizabeth, see Montrose, "*A Midsummer Night's Dream* and the Shaping Fantasies of Elizabethan Culture: Gender, Power, Form." On other matters of gender and metamorphosis in the play,
see Clarke, *Allegories of Writing*, 128–32.

28. Shakespeare, *Midsummer*, 2.1.156–57.

29. Ibid., 2.1.128–29.

30. Ibid., 5.1.1–27.

31. Von Foerster translates this insight into systems discourse in the remark
that "reality appears as a consistent reference frame for at least two observers."
See "On Self-Organizing Systems," *Understanding Understanding*, 4. See also
von Foerster's essay "Objects: Tokens for (Eigen-) Behaviors," *Understanding
Understanding*, 261–71.

32. Shakespeare, *Midsummer*, 4.1.214.

33. Ibid., 1.1.150, 153.

34. Ibid., 1.1.145–48.

35. Ibid., 1.1.250–51.

36. Ibid., 5.1.12–17.

37. Redrawn from Manfred Jahn's web text, *Narratology: A Guide to the Theory of Narrative*. For an application of Bremond's system to fabula construction,
see Bal, *Narratology*, 2nd. ed., 188–93.

38. Genette, *Narrative Discourse*, 228

39. Stanzel, *Theory of Narrative*, 4.

40. Pynchon, *The Crying of Lot 49*, 124.

41. Ibid., 124–25.

42. One must either construct a particular interpretation, as I have, or let
the matter oscillate without resolution. In the Pynchon passage, for instance,
one cannot be finally sure whether the thoughts being set forth are Oedipa's or
the narrator's. In "Systems Theory and the Difference between Communication and Consciousness: An Introduction to a Problem and its Context," Dietrich Schwanitz points out the communicational problematics of attribution in

the figural narrative situation: "Another paradox of communication can be found in the novel with an impersonal narrator. Here, the crucial matter is the uncertainty of attribution of the narration as a communicative act, which can be referred both to the character experiencing the events and to the invisible narrator. This is expressed in the narrative technique of combining both perspectives in free indirect discourse" (496–97).

43. See Jakobson, "Linguistics and Poetics."

44. Von Foerster has investigated how it is possible to increase the order of a system, not only by decreasing its entropy, but also by maintaining its entropy while increasing its complexity: see his "On Self-Organizing Systems," 7–9, and my "Heinz von Foerster's Demons: Self-Organization and the Emergence of Second-Order Systems Theory," in *Emergence and Embodiment: New Essays in Second-Order Systems Theory*, ed. Clarke.

45. See Nicholas Royle, *Telepathy and Literature: Essays on the Reading Mind.*

46. See Luhmann, "Paradox": "Tradition teaches us to conceptualize form as one side of a distinction whose other side then can be designated in various ways" (15); section II of Luhmann, "Identity—What or How?" See also Esposito, "Two-Sided Forms in Language."

47. For instance, Bois, *Formless: A User's Guide.*

48. Luhmann, *Art*, 28

49. Luhmann, "Paradox," 87.

50. Wellbery, "Redescription: Literary Semiotics, Deconstruction, Systems Theory," italics in original.

51. Knight, *Beyond the Barrier.*

52. Vertical or ontological embedding is further defined and discussed in the "Bateson's Play Frame" section of chapter 4.

53. Knight, *Beyond the Barrier*, 151–52.

54. I discovered this forgotten novel in Smithson, "Entropy and the New Monuments." The larger passage reads: "As action decreases, the clarity of such surface structures increases. This is evident when all representations of action pass into oblivion. At this stage, lethargy is elevated to the most glorious magnitude. In Damon Knight's Sci-fic novel, 'Beyond the Barrier,' he describes in a phenomenological manner just such surface-structures: 'Part of the scene before them seemed to expand'" (14). Smithson's delving into science-fiction surface-structure retrieves textual detail for observation by an artist's eye attuned to the depths within surfaces, while also nodding toward dark science fiction because it puts the human race in its cosmic place. Thermodynamic entropy is the dark side of systemic transformation. But, to be precise, thermodynamic entropy produces maximal dissipation specifically in systems *environmentally* closed to energy input. The autopoietic systems we are focused on are *operationally* closed but environmentally open.

55. Knight, *Beyond the Barrier*, 182.

56. Ibid., 186.

CHAPTER 4: METAMORPHOSIS AND EMBEDDING

1. Barth, "Frame-Tale" and "Menelaiad"; and "Tales Within Tales Within Tales."

2. For more on this episode of the *Odyssey*, see Clarke, *Allegories*, 58–60.

3. Nelles, *Frameworks: Narrative Levels and Embedded Narrative*, 150.

4. Borges, "Partial Magic in the *Quixote*," 194. Borges's discussion of Cervantes' and other classic authors' forms of circular play with narrative frames precludes the notion that postmodernist fictions have any monopoly on them. Standard references to metafiction are Hutcheon, *Narcissistic Narrative: The Metafictional Paradox*; and Waugh, *Metafiction: The Theory and Practice of Self-Conscious Fiction*. See also Pearce, "Enter the Frame"; and Ryan, "Metaleptic Machines" in *Avatars*, 204–30.

5. Genette, *Narrative Discourse*, 234–35.

6. Borges, "*Quixote*," 196.

7. Todorov, "Narrative-Men," in *Poetics*, 72–73. Calvino ties together Borges and Todorov in his 1978 paper "Levels of Reality in Literature," in *Uses*, 101–21.

8. Bal also stresses this distinction in no uncertain terms: "Character is intuitively the most crucial category of narrative, and also most subject to projection and fallacies"; to avoid these interpretive pitfalls, narratology must take an analytical view of literary character as "an enactment of radical constructivism: a character is a construction, not a person" (*Narratology*, 115, 122).

9. Todorov, *Poetics*, 67.

10. Ibid., 73.

11. Ibid., 76.

12. Ibid., 78.

13. Bal, *Narratology*, 53.

14. Ibid., 54–55.

15. Ibid., 57.

16. For more on narrative framing in *Frankenstein*, see Beth Newman, "Narratives of Seduction and the Seductions of Narrative: The Frame Structure of *Frankenstein*."

17. Nelles, *Frameworks*, 139.

18. On the tale of Cupid and Psyche and its positioning within *The Golden Ass*, see Clarke, *Allegories*, 115–22. In this study, mesmerized by the thematics of allegory, I wrote at length on the embedded structures of this and other stories of metamorphic changes without ever seizing embedded narration as a topic in its own right. I can trace my critical awakening on this score precisely to Jim Paxson's discussion of narrative "containment" in "Kepler's Allegory of Containment, the Making of Modern Astronomy, and the Semiotics of Mathematical Thought."

19. Nelles, *Frameworks*, 139–40.

20. Ibid., 149.

21. Ibid.

22. Gregory Bateson, "A Theory of Play and Fantasy," *Steps*, 177–93.

23. Ibid., 177–78.

24. Ibid., 178.

25. Ibid.

26. Ibid., 179–80.

27. For another exposition of Bateson's play frame, as well as an important document marking the emergence of the neocybernetic paradigm in relation to structural psychoanalysis, see Wilden, *System and Structure*, esp. 17–18. See also Wolfe, *Zoontologies*, 39–41.

28. Bateson, *Steps*, 188.

29. Todorov, *Poetics*, 76.

30. Ibid., 76–77.

31. Calvino, *Uses of Literature*, 115–16. Cf. Dirk Baecker on the function of the "unwritten cross" in the calculus of form, in "The Form Game."

32. Derrida, *The Truth in Painting*, 59.

33. The discourse of the *pharmakon* is developed in Derrida, *Plato's Pharmacy*. In *The Truth in Painting*, Derrida's immediate example of a conceptual frame is Kant's use of a logical structure extracted from the *Critique of Pure Reason* to organize the Analytic of the Beautiful in the *Critique of Judgment*. The arguable inappropriateness of the frame in question only compounds the parergonal paradoxicality of Kant's aesthetic formula for beauty as the subjective experience of an object possessing "purposiveness without purpose" [*Zweckmässigkeit ohne Zweck*].

34. Derrida, *Truth*, 61.

35. For more on the interplay between deconstruction and systems theory, see Wolfe, "Meaning as Event-Machine, or, Systems Theory and 'The Reconstruction of Deconstruction.'"

36. Lem, *The Cyberiad: Fables for the Cybernetic Age*. A foundational discussion for my approach is Hayles, "Chaos as Dialectic: Stanislaw Lem and the Space of Writing."

37. Lem, *Cyberiad*, 167.

38. Von Neumann, "The General and Logical Theory of Automata," 315.

39. Von Neumann, "Automata," 315–16.

40. Lem, *Cyberiad*, 115.

41. Ibid., 116.

42. Ashby, *Introduction to Cybernetics*, 10.

43. Ibid., 11.

44. Ibid.

45. *See* Ashby, "Coupling Systems," in *Introduction to Cybernetics*, 48ff.

46. Lem, *Cyberiad*, 196.

47. Ibid., 195.

48. Ibid.

49. Ibid., 201–2.

50. Stites, *Revolutionary Dreams: Utopian Vision and Experimental Life in the Russian Revolution*, see especially chapter 7, "Man the Machine" (145–64). In a way that anticipated the Soviet vogue for cybernetics, the capitalist efficiency craze of the early twentieth century also leant itself directly to revolutionary movements for economic collectivization. These ironies of history are an important backdrop to the fate of the first cybernetics, which offered computerization itself as the latest in the systemization of work and control of information. As Margaret Mead remembers, with his books on cybernetics, Norbert Wiener "went to Russia, and was very very well received. The Russians were crazy about this right away—it fit right into their lives. . . . Cybernetics spread all over the Soviet Union very rapidly" (Brand, "For God's Sake, Margaret," 37). See Gerovitch, *From Newspeak to Cyberspeak: A History of Soviet Cybernetics*.

51. Lem, *Cyberiad*, 209.

52. See Lem's detailed account of "phantomatics," a fictive virtual-reality technology he first concocted in 1961–63 and presented in *Summa Technologiae*, which "futurological" speculations he revisits in the wake of the virtual-reality craze and emergence of actual virtual-reality technology, in "Thirty Years Later."

53. Lem, *Cyberiad*, 224.

54. Ibid., 225.

55. Ibid., 226–27.

56. Ibid., 227.

57. Ibid., 228, 229.

58. Ibid., 229–30.

59. Ibid., 44.

60. Ibid., 45.

61. Ibid., 173.

62. Ibid., 195.

63. Darwin, *Origin of Species*, 174.

64. For a critique of the late-humanist hangover imposed upon Darwin by Darwinist discourse, see Gray, *Straw Dogs: Thoughts on Humans and Other Animals*.

65. Jaron Lanier criticizes the breed of popular posthumanism he calls "cybernetic totalism" in "One Half of a Manifesto."

66. This version is given a second, stronger spin in the novel's parting episode, "Prince Ferrix and the Princess Crystal," a Swiftean satire of human sexuality as an abomination: humanity "originated in a manner as mysterious as it was obscene. There arose noxious exhalations and putrid excrescences, and out of these was spawned the species known as paleface—though not all at once. First, there were creeping molds that slithered forth. . . . Our progenitor created the algorithm of electroincarnation and in the sweat of his brow begat our kind,

thus delivering machines from the house of paleface bondage" (Lem, *Cyberiad*, 283–84).

67. Ibid., 200.

68. Ibid., 244–45.

69. Ashby, *Cybernetics*, 1. Ashby's "The materiality is irrelevant" was an unfortunately elliptical way of stating that the given nature of the materiality is less important than its virtual availability. In our world, however, some medium upon which to impress or through which to transmit the informatic pattern is indispensable. Ashby does not actually imply that substantial instantiations such as bodies are expendable, but rather, that they are not the operationally determinative factors for the full range of the "behaviors" of the systems brought into being through the in-forming of their material elements.

70. Bateson, *Steps*, 401.

71. Ibid., xxiv–xxv.

72. Coveney, *Frontiers of Complexity*, 17.

73. Luhmann, "Control of Intransparency," 361.

74. Lem, *Cyberiad*, 260.

75. Ibid., 263.

76. Ashby, *Cybernetics*, 9.

CHAPTER 5: COMMUNICATING *THE FLY*

1. Kittler, *Gramaphone*, 28.

2. "Communication *is* the creation of redundancy or patterning" (Bateson, *Steps*, 406).

3. Kittler's hermeneutics of the hardware stands behind Hans Ulrich Gumbrecht and Timothy Lenoir's discussions of "the materialities of communication"—the importance of determining the historical particularities of the instruments and networks that enable communications media to function at all. See Gumbrecht, ed., *Materialities of Communication*; and Lenoir, ed., *Inscribing Science: Scientific Texts and the Materialities of Communication*, esp. Lenoir's introductory essay "Inscription Practices and the Materialities of Communication," 1–19.

4. Luhmann, *Social Systems*, 139. Luhmann touches on the concept of noise soon after: "The combination of information, utterance, and expectation of success in one act of attention presupposes 'coding.' The utterance must duplicate the information, that is, on the one hand, leave it outside yet, on the other, use it for utterance and reformulate it appropriately: for example, by providing it with a linguistic (eventually an acoustic, written, etc.) form. . . . What is sociologically important is, above all, that this too brings about a differentiation within the communication process. Events must be distinguished as coded and uncoded. Coded events operate as information in the communication process, uncoded ones as disturbance (noise)" (142).

5. Redrawn from Weaver, "Recent Contributions to the Mathematical Theory of Communication," 7.

6. Paulson, *Noise*, 67.

7. See Weaver, "Recent Contributions," 18–22. On classical thermodynamics, see Clarke, *Energy Forms: Allegory and Science in the Era of Classical Thermodynamics*.

8. Kittler, *Gramaphone*, 45.

9. Serres, *Hermes*, 67. See also Serres, *The Parasite*: "The noise, through its presence and absence, the intermittence of the signal, produces the new system" (52).

10. Serres, *Hermes*, 77.

11. Atlan, "Hierarchical Self-Organization in Living Systems: Noise and Meaning," 196.

12. Langelaan, "The Fly," *Playboy*. In addition to occasional short stories, his literary output is a memoir of undercover work for the British Secret Service during World War II. To operate incognito in Nazi-occupied France, he submitted to plastic surgery to change his face, especially his ears and chin. See Langelaan, *The Masks of War*, 81–89.

13. *The Fly*, dir. Neumann. I treat this film more extensively in "Mediating *The Fly*: Posthuman Metamorphosis in the 1950s."

14. *The Fly*, dir. Cronenberg. For information on many textual and production details on Cronenberg's *Fly*, see Kirkman, "The Annotated Fly." See also Pharr, "From Pathos to Tragedy: The Two Versions of *The Fly*"; Roth, "Twice Two: *The Fly* and *Invasion of the Body Snatchers*"; Knee, "The Metamorphosis of *The Fly*"; Freeland, "Feminist Frameworks for Horror Films"; Wicke, "Fin de Siècle and the Technological Sublime"; and Telotte, "Crossing Genre Boundaries/Bound by Fantasy."

15. Langelaan, "The Fly," *The Playboy Book of Science Fiction and Fantasy*, 24; all further citations are from this volume. The text is also available in *Reel Future*, ed. Ackerman.

16. Bal, *Narratology*, 137.

17. To focus on *The Fly* solely as a castration scenario would be to ignore the specifics of the cybernetic framework. In *How We Became Posthuman*, Katherine Hayles notes that "changes in bodies as they are represented within literary texts have deep connections with changes in textual bodies as they are encoded within information media. . . . The contemporary pressure toward dematerialization, understood as an epistemic shift toward pattern/randomness and away from presence/absence, affects human and textual bodies on two levels at once, as a change in the body (the material substrate) and a change in the message (the codes of representation)" (29). Hayles defines this change as a shift from *castration* to *mutation*. "Mutation is crucial, because it names the bifurcation point at which the interplay between pattern and randomness causes the system to evolve in a new direction" (33).

18. Genette's definition of *pseudo-diegesis* is "telling as if it were diegetic (as if it were at the same narrative level as its context) something that has nevertheless been presented as (or can easily be guessed to be) metadiegetic in its principle or, if one prefers, in its origin" (*Narrative Discourse*, 236); more simply put, it is embedded narration rendered as if it were *not* embedded. David Alan Black explains: "The story-within-the-story, in other words, becomes the story. One generation of quotation marks is excised; everything beyond it shifts by one level" ("Genette and Film: Narrative Level in the Fiction Cinema," 22). The sort of unaccountable surplus that attributes to Hélène's narrated information a perception she could not have experienced is what Genette terms *paralepsis*: "giving more than is authorized in principle by the code of focalization governing the whole" (*Narrative Discourse*, 195).

19. This play of embeddings at the intersection of mediation and embodiment announces a set of problematics that only become more severe with the development of digital media platforms. Mark B. N. Hansen investigates these issues at length in "Cinema Beyond Cybernetics, or How to Frame the Digital Image."

20. *The Fly*, dir. Neumann.

21. Physicist Anton Zeilinger notes that "the procedure for teleportation in science fiction . . . generally goes as follows: A device scans the original object to extract all the information needed to describe it. A transmitter sends the information to the receiver station, where it is used to obtain an exact replica of the original. In some cases, the material that made up the original is also transported to the receiving station, perhaps as 'energy' of some kind; in other cases, the replica is made of atoms and molecules that were already present at the receiving station" ("Quantum Teleportation," 50).

22. Wiener, *The Human Use of Human Beings: Cybernetics and Society*, 105–11.

23. Luhmann, *Social Systems*, 139.

24. See Hendershot, "The Cold War Horror Film: Taboo and Transgression in *The Bad Seed*, *The Fly*, and *Psycho*."

25. Langelaan, "The Fly," 17–18.

26. Ibid., 20.

27. Ibid., 20

28. The interposition of a fly between the sender and receiver of a "ghostly" message was imagined a century earlier by Emily Dickinson: "and then it was / There interposed a Fly— / With Blue—uncertain stumbling Buzz— / Between the light—and me" (#465).

29. Langelaan, "The Fly," 21.

30. Von Foerster, "Perception of the Future and the Future of Perception," 208.

31. Langelaan, "The Fly," 35–36.

32. Von Foerster, "Perception," 208.

33. Wicke, "Fin de Siècle," 305.

34. *The Fly*, dir. Cronenberg.

35. Ibid.

36. Langelaan, "The Fly," 23.

37. *The Fly*, dir. Cronenberg.

38. Ibid.

39. Ibid.

40. See Clarke, *Allegories*, esp. 122–28.

41. As we noted in chapter 4, "The constructing automaton is supposed to be placed in a reservoir in which all elementary components are floating, and it will effect its construction in that milieu" (von Neumann, "Automata," 316).

42. See Deleuze, "1730: Becoming-Intense, Becoming-Animal, Becoming Imperceptible . . . ," ch.10 of *A Thousand Plateaus: Capitalism and Schizophrenia*, for an approach to metamorphosis as deterritorialization. Braidotti gives Deleuzian readings of becoming-woman and becoming-machine in Neumann's and Cronenberg's *Fly*s in *Metamorphoses: Towards a Materialist Theory of Becoming*.

43. Wicke discusses the Christ-likeness of Seth Brundle in "Fin de Siècle," 306–7.

44. *The Fly*, dir. Cronenberg.

45. Helen W. Robbins has written well on the theme of womb envy in this film and Cronenberg's following production, *Dead Ringers*, in "'More Human Than I Am Alone': Womb Envy in David Cronenberg's *The Fly* and *Dead Ringers*."

46. See Clarke, *Allegories*, ch.5, "The Gender of Metamorphosis," 113–47.

47. *The Fly*, dir. Cronenberg.

48. Ibid.

49. Ibid.

50. Ibid.

51. Pay, ed., "Dreams." A shorter version of this passage is offered as an example of paradox in Hughes, *Vicious Circles*, 35.

52. Kafka, *Metamorphosis*.

53. *The Fly*, dir. Cronenberg.

54. Latour, *Aramis*, 24, 57.

55. *The Fly*, dir. Cronenberg.

56. On the "space bug," see Kirkman, "Annotated Fly."

57. *The Fly*, dir. Cronenberg.

58. Margulis, "Gaia is a Tough Bitch," 134. For an in-depth presentation of symbiogenesis, see Margulis, *Acquiring Genomes: A Theory of the Origins of Species*.

59. *The Fly*, dir. Cronenberg.

60. See Foster, "'The Sex Appeal of the Inorganic': Posthuman Narratives and the Construction of Desire."

CHAPTER 6: POSTHUMAN VIABILITY

1. Sterling, *Schismatrix*, 170. For a discussion stressing more the post- than the humanism of this novel, see Bukatman, "Postcards from the Posthuman Solar System."

2. Sterling, *Schismatrix*, 114.

3. Ibid., 232.

4. See for instance Bonner, "Difference and Desire, Slavery and Seduction: Octavia Butler's *Xenogenesis*"; Boulter, "Polymorphous Futures: Octavia E. Butler's *Xenogenesis* Trilogy"; Alaimo, "Playing Nature: Postmodern Natures in Contemporary Feminist Fiction"; Michaels, "Political Science Fictions"; Mehaffy, "'Radio Imagination': Octavia Butler on the Poetics of Narrative Embodiment"; and Brataas, "Becoming Utopia in Octavia E. Butler's *Xenogenesis* Series." The turn toward a critique of the narrative invoking the posthuman is well handled in Jacobs, "Posthuman Bodies and Agency in Octavia Butler's *Xenogenesis*."

5. For Butler's own distinction between science fiction and fantasy, see Rowell, "An Interview with Octavia E. Butler."

6. Butler, *Parable of the Talents*, 46.

7. Butler's first five novels (apart from *Kindred*) are interconnected by a shifting set of shared characters: *Patternmaster* (1976), *Mind of My Mind* (1977), *Survivor* (1978), *Wild Seed* (1980), and *Clay's Ark* (1984).

8. See for instance Smith, "Morphing, Materialism, and the Marketing of *Xenogenesis*"; Holden, "The High Costs of Cyborg Survival: Octavia Butler's *Xenogenesis* Trilogy"; and, in Korean, Yu, "The Representation of Inappropriate/d Others: The Epistemology of Donna Haraway's Cyborg Feminism and Octavia Butler's *Xenogenesis* Series."

9. Haraway, "Cyborg Manifesto," 179. Haraway's other treatments of or references to Octavia Butler occur in "Reprise: Science Fiction, Fictions of Science, and Primatology," in *Primate Visions*, 368–82 "The Biopolitics of Postmodern Bodies: Constitutions of Self in Immune System Discourse," in *Simians*, 203–30; and *Modest_Witness*, 229.

10. Haraway, *Modest_Witness*, 14.

11. Ibid., 8–9.

12. Butler, *Lilith's Brood*, 542. All quotations from Butler's Xenogenesis trilogy will be taken from *Lilith's Brood*, the 2000 omnibus edition incorporating *Dawn*, *Adulthood Rites*, and *Imago*.

13. Haraway, *Modest_Witness*, 108.

14. Ibid., 56.

15. Ibid., 52.

16. Butler, *Lilith's Brood*, 283.

17. See the chapter "Universal Donors in a Vampire Culture: It's All in the Family. Biological Kinship Categories in the Twentieth-Century United States," in Haraway, *Modest_Witness*, 213–65.

18. Haraway, *Modest_Witness*, 362.

19. On the ambivalence marked by the demonic–daemonic distinction, see Clarke, *Allegories*, 12–13.

20. Butler, *Wild Seed*, 15.

21. Ibid., 86–87.

22. Slonczewski, "Octavia Butler's *Xenogenesis* Trilogy: A Biologist's Response." This text makes no mention of Lynn Margulis.

23. Cathy Peppers, "Dialogic Origins and Alien Identities in Butler's Xenogenesis."

24. Ibid., 60.

25. Max, "The Literary Darwinists," 77. See Clarke, "Charles Darwin: Conservative Messiah? On Joseph Carroll's *Literary Darwinism*."

26. Peppers, "Dialogic Origins," 54. The Margulis and Sagan excerpt Peppers refers to is prefaced by science writer Jeanne McDermott's appreciation, "Lynn Margulis: Vindicated Heretic." Most recently, Luciana Parisi has indicated the Margulis–Butler connection in her remarkable *Abstract Sex: Philosophy, Bio-Technology and the Mutations of Desire*, which links the discourse of Margulis and Sagan with Haraway's cyborg writings and the biophilosophy of Deleuze and Guattari to develop a speculative itinerary for the unfolding of posthuman sexuality toward a "biodigital" stratum. When discussing Margulis and Sagan, Parisi occasionally sets in without commentary epigraphs from *Dawn* and *Imago*, suggesting that in the current milieu of high-theoretical cyberfeminism, Butler's work has achieved a kind of iconic status that goes without saying. See also Parisi, "Essence and Virtuality: The Incorporeal Desire of Lilith."

27. Personal communication by e-mail received August 21, 2005.

28. See Dick, *The Living Universe: NASA and the Development of Astrobiology*, 48.

29. Margulis, *Microcosmos: Four Billion Years of Evolution from Our Microbial Ancestors* (New York: Summit Books, 1986). Unless otherwise noted, all citations from *Microcosmos* are taken from this edition.

30. Margulis, *Microcosmos: Four Billion Years of Microbial Evolution*.

31. Ibid., 13–14.

32. Butler, *Lilith's Brood*, 8.

33. Margulis, *Microcosmos*, 14–15.

34. Butler, *Lilith's Brood*, 40.

35. Margulis, *Microcosmos*, 15.

36. Ibid., 16.

37. Ibid., 82.

38. Ibid., 85.

39. In *Acquiring Genomes*, Margulis and Sagan note Donald I. Williamson's argument "that the major evolutionary changes in backboneless animals (and nearly all animals are invertebrates) emanate from inheritance of acquired genomes. He argues that the genomes that determine the larval animal forms are different from those that determine the adult forms. . . . They are from other animals"; they cite Williamson: "'the methods of metamorphosis which link successive phases in development could not have evolved merely by natural selection of random mutations. . . . [While] adults and larvae have evolved by "descent with modification," . . . superimposed on this process, entire genomes

have been transferred by hybridization'" (165–66). See Williamson, *The Origins of Larvae*.

40. Butler, *Lilith's Brood*, 543.
41. Margulis, *Microcosmos*, 22.
42. Butler, *Lilith's Brood*, 55.
43. Ibid., 75.
44. Ibid., 15, 116.
45. Ibid., 36.
46. Margulis, *Microcosmos*, 20.
47. See Otis, *Organic Memory: History and the Body in the Late Nineteenth & Early Twentieth Centuries*: "The nineteenth-century organic memory theory . . . proposed that memory and heredity were essentially the same and that one inherited memories from ancestors along with their physical features . . . placed the past *in* the individual, *in* the body, *in* the nervous system; it pulled memory from the domain of the metaphysical into the domain of the physical with the intention of making it *knowable*. . . . Repeated patterns of sensations, whether of the recent or distant past, had left traces in the body, making the individual an epitome of his or her racial history" (2–3).
48. Margulis, *Microcosmos*, 93
49. Butler, *Lilith's Brood*, 63.
50. Margulis, *Microcosmos*, 20.
51. Butler, *Lilith's Brood*, 63.
52. For a more recent account of these matters, entirely congruent with Margulis and Sagan's work, see Sonea, *Prokaryotology: A Coherent View*.
53. Margulis, *Microcosmos*, 119.
54. Ibid., 17.
55. Butler, *Lilith's Brood*, 41.
56. See White, "The Erotics of Becoming: *Xenogenesis* and 'The Thing'": "To the extent that the Oankali have a specific nature distinguishing them from other life-forms, this 'origin' founding their identity as a distinct group would reside in that 'miniscule cell within a cell,' a life-form so rudimentary it amounts to little more than a genetically-encoded instruction to become other" (403).
57. Margulis, *Microcosmos*, 117.
58. Ibid., 126.
59. Ibid., 196.
60. Butler, *Lilith's Brood*, 60–61.
61. Ibid., 73.
62. Ibid., 75.
63. Ibid., 79.
64. Ibid., 84.
65. Ibid., 106.
66. Ibid., 107.
67. Margulis, *Microcosmos*, 81–82.

68. Butler, *Lilith's Brood*, 109.
69. Ibid., 543.
70. Ibid., 693.
71. Ibid., 680.
72. Ibid., 618.
73. Ibid., 162.
74. Ibid.
75. Ibid., 646.
76. Ibid., 22.
77. Ibid., 31.
78. Ibid., 41.
79. Ibid., 452–53.
80. Ibid., 237.
81. Ibid., 156.
82. Ibid., 253.
83. Ibid., 352.
84. Ibid., 496.
85. Ibid., 547.
86. One suspects that here Butler is doing a riff on neoteny, "the retention of youthful traits into adulthood" (Margulis, *Microcosmos*, 208), a developmental dynamic whose contributions to the evolution of *Homo sapiens* Margulis and Sagan speculate upon in detail: "We are exposed and molded by the influences of the outside world while we are still malleable and vulnerable" (209).
87. Ibid., 524.
88. Ibid., 538.
89. Ibid., 536, 542.
90. Ibid., 540.
91. Ibid., 545.
92. Ibid., 541.
93. Ibid., 283.
94. Ibid., 554.
95. Ibid., 591.
96. Ibid., 607.
97. Ibid., 610–11.
98. Ibid., 612.
99. Ibid., 615.
100. Ibid., 628.
101. Ibid., 637.
102. Ibid., 630.
103. Ibid., 647.
104. Ibid., 674.
105. White, "Erotics of Becoming," 406.
106. Butler, *Lilith's Brood*, 683, 681.

107. Ibid., 682.

108. Margulis, *Microcosmos*, 148.

109. Butler, *Lilith's Brood*, 685.

110. The phrase could be a typo for "Genetic memories," but it reads "Genetics" both in *Lilith's Brood* and in the last freestanding edition of *Imago* (New York: Warner Books, 1997), 167. Granting that this is the text Butler intended, it would imply that these are memories not just of the dispositions of genes but of the genetic knowledge or biological discipline the Oankali have built up of, and on the basis of, their ongoing accumulation and deliberate manipulation of genes. That is, this phrasing stresses the science of genetics, for which the Oankali are the extraterrestrial allegory.

111. Butler, *Lilith's Brood*, 692–93.

112. Keats, *Hyperion: A Fragment*, in *John Keats: Complete Poems*, 268.

113. Whitman, "Out of the Cradle Endlessly Rocking," in *Leaves of Grass and Selected Poems*, 202.

114. Butler, *Lilith's Brood,*, 742.

115. See Weiskel, *The Romantic Sublime: Studies in the Structure and Psychology of Transcendence*. Romantic sublimity is given a cogent systems–theoretical redescription in Economides, "'Mont Blanc.'"

116. Butler, *Lilith's Brood*, 744–45.

117. See "The Future Supercosm," chapter 13 of Margulis, *Microcosmos*, 235–75.

118. Butler, *Lilith's Brood*, 746.

CONCLUSION: THE NEOCYBERNETIC POSTHUMAN

1. "Garish predictions": see for instance, Kurzweil, *The Age of Spiritual Machines: When Computers Exceed Human Intelligence*.

2. Luhmann, *Social Systems*, 35.

3. Ibid., 39–40.

4. Sagan, "The Uncut Self," 22, my italics.

5. Luhmann, *Art*, 10.

Abbott, H. Porter. *The Cambridge Introduction to Narrative*. New York: Cambridge University Press, 2002.

Alaimo, Stacy. "Playing Nature: Postmodern Natures in Contemporary Feminist Fiction." In *Undomesticated Ground: Recasting Nature as Feminist Space*, by Alaimo. Ithaca, N.Y.: Cornell University Press, 2000. 133–70.

Ashby, W. Ross. *An Introduction to Cybernetics*, 2nd. ed. London: Chapman and Hall, 1956. Available at http://pcp.vub.ac.be/books/IntroCyb.pdf.

Atlan, Henri. "Hierarchical Self-Organization in Living Systems: Noise and Meaning." In *Autopoiesis: A Theory of Living Organization*, edited by Milan Zeleny. New York: Elsevier North Holland, 1981. 185–208.

Badmington, Neil, ed. "Posthuman Conditions." Special issue, *Subject Matters* 3, no. 2, 4, no. 1 (2007).

———, ed. *Posthumanism*. New York: Palgrave, 2000.

Baecker, Dirk. "The Form Game." In *Problems of Form*, edited by Baecker, 99–106.

———, ed. *Problems of Form*. Stanford, Calif.: Stanford University Press, 1999.

Bal, Mieke. *Narratology: Introduction to the Theory of Narrative*, 2nd ed. Buffalo: University of Toronto Press, 1997.

Barlow, Connie, ed. *From Gaia to Selfish Genes: Selected Writings in the Life Sciences*. Cambridge, Mass.: MIT Press, 1991.

Barth, John. *The Friday Book: Essays and Other Nonfiction*. New York: Putnam, 1984.

———. *Lost in the Funhouse*. New York: Bantam, 1969.

Bateson, Gregory. *Steps to an Ecology of Mind*. New York: Ballantine, 1972.

Bendle, Mervyn F. "Teleportation, Cyborgs and the Posthuman Ideology." In *Social Semiotics* 12, no. 1 (2002): 45–62

Berreby, David. ". . . that damned elusive Bruno Latour." In *Lingua Franca* (September/October 1994): 22–32, 78.

Bevington, David, ed. *The Complete Works of Shakespeare*, 5th ed. New York: Longman, 2003.

Bjerg, Ole. "Accelerating Luhmann: Towards a Systems Theory of Ambivalence." In *Theory, Culture and Society* 23, no. 5 (2006): 49–68.

Black, David Alan. "Genette and Film: Narrative Level in the Fiction Cinema." In *Wide Angle* 8, nos. 3–4 (1986): 19–26.

Bois, Yves-Alain and Rosalind Krauss. *Formless: A User's Guide*. New York: Zone, 1997.

Bolter, Jay and Richard Grusin. *Remediation: Understanding New Media*. Cambridge, Mass.: MIT Press, 1999.

Bonner, Frances. "Difference and Desire, Slavery and Seduction: Octavia Butler's *Xenogenesis*." In *Foundation* 48 (Spring 1990): 50–62.

Borges, Jorge Luis. *Ficciones*. Ed. Anthony Kerrigan. New York: Grove Wiedenfeld, 1963.

———. *Labyrinths: Selected Stories and Other Writings*. Ed. Donald A. Yates and James E. Irby. New York: New Directions, 1964.

———. *Other Inquisitions 1937–1952*. Trans. Ruth L. C. Simms. Austin: University of Texas Press, 1964.

Boulter, Amanda. "Polymorphous Futures: Octavia E. Butler's *Xenogenesis* Trilogy." In *American Bodies: Cultural Histories of the Physique*, edited by Tim Armstrong. New York: New York University Press. 170–85.

Braidotti, Rosi. *Metamorphoses: Towards a Materialist Theory of Becoming*. Cambridge, Mass.: Polity Press, 2002.

———. "Posthuman, All Too Human: Towards a New Process Ontology." In *Theory, Culture and Society* 23, nos.7–8 (2006): 197–208.

Brand, Stewart. "For God's Sake, Margaret: Conversation with Gregory Bateson and Margaret Mead." In *The CoEvolution Quarterly* 10 (Summer 1976): 32–44.

Brataas, Delilah Bermudez. "Becoming Utopia in Octavia E. Butler's *Xenogenesis* Series." In *Foundation: The International Review of Science Fiction* 35, no. 96 (Spring 2006). 84–100.

Brier, Søren. "The Construction of Information and Communication: A Cybersemiotic Reentry into Heinz von Foerster's Metaphysical Construction of Second-Order Cybernetics." In *Semiotica* 154, nos. 1–4 (2005): 355–99.

Bukatman, Scott. "Postcards from the Posthuman Solar System." In *Science-Fiction Studies* 18, no. 3 (1991): 343–57.

Butler, Octavia. *Imago*. 1989. New York: Warner, 1997.

———. *Lilith's Brood*. New York: Warner, 2000.

———. *Parable of the Talents*. New York: Warner, 1998.

———. *Wild Seed*. 1980. New York: Warner, 1999.

Calvino, Italo. *The Uses of Literature: Essays*. Trans. Patrick Creagh. New York: Harcourt Brace Jovanovich, 1986.

Clarke, Arthur C. *Childhood's End*. New York: Del Rey, 1990.

Clarke, Bruce. *Allegories of Writing: The Subject of Metamorphosis*. Albany: State University of New York Press, 1995.

———. "Charles Darwin: Conservative Messiah? On Joseph Carroll's *Literary Darwinism*." In *Green Letters* 7 (Spring 2006): 36–41.

———. *Energy Forms: Allegory and Science in the Era of Classical Thermodynamics.* Ann Arbor: University of Michigan Press, 2001.

———. "Friedrich Kittler's Technosublime." In *electronic book review* 10 (Winter 1999–2000) http://www.altx.com/ebr/reviews/rev10/r10cla.htm; rp. *American Book Review* 22, no. 1 (November/December 2000): 7, 9.

———. "Mediating *The Fly*: Posthuman Metamorphosis in the 1950s." In *Configurations* 10, no. 1 (Winter 2002): 169–91.

———. "Science, Theory, and Systems." In *Interdisciplinary Studies in Literature and Environment* 8, no. 1 (Winter 2001): 149–65.

———. "Strong Constructivism: Modernity and Complexity in Science Studies and Systems Theory." In *Democracy, Civil Society, and Environment*, edited by Joseph Bilello. Muncie, Ind.: College of Architecture and Planning Monograph, Ball State University, 2002. 41–49.

Clarke, Bruce, and Mark B. Hansen, eds. *Emergence and Embodiment: New Essays in Second-Order Systems Theory*. Durham, N.C.: Duke University Press, 2009.

Clarke, Bruce, and Linda D. Henderson, eds. *From Energy to Information: Representation in Science and Technology, Art, and Literature*. Stanford, Calif.: Stanford University Press, 2002.

Cobley, Paul. *Narrative*. New York: Routledge, 2001.

———. "Narratology." In *The Johns Hopkins Guide to Literary Theory and Criticism*, 2nd ed., edited by Michael Groden, Martin Kreiswirth, and Imre Szeman. Baltimore: Johns Hopkins University Press, 2005. 677–82.

Cornell, Drucilla. "Enabling Paradoxes: Gender Difference and Systems Theory." In *NLH* 27, no. 2 (Spring 1996): 185–97.

Coveney, Peter, and Roger Highfield. *Frontiers of Complexity: The Search for Order in a Chaotic World*. New York: Fawcett Columbine, 1995.

Crawford, T. Hugh. "Imaging the Human Body: Quasi Objects, Quasi Texts, and the Theater of Proof." In *PMLA* 111, no. 1 (January 1996): 66–79.

———. "Networking the (Non)Human: *Moby-Dick*, Matthew Fontaine Maury, and Bruno Latour." In *Configurations* 5, no. 1 (Winter 1997): 1–21.

Culler, Jonathan. *The Pursuit of Signs: Semiotics, Literature, Deconstruction*. Ithaca, N.Y.: Cornell University Press, 1981.

Darwin, Charles. *Darwin: Texts, Commentary*. Ed. Philip Appleman. 3rd ed. New York: Norton, 2001.

de Berg, Henk. "Luhmann's Systems Theory and Its Applications in Literary Studies: A Bibliography." In *European Journal of English Studies* 5, no. 3 (December 2001): 385–436.

Deleuze, Gilles and Felix Guattari. *A Thousand Plateaus: Capitalism and Schizophrenia*. Trans. Brian Massumi. Minneapolis: University of Minnesota Press, 1987.

Derrida, Jacques. *Plato's Pharmacy*. In *Dissemination*, by Jacques Derrida, translated by Barbara Johnson. Chicago: University of Chicago Press, 1981. 63–169.

———. *The Truth in Painting*. Translated by Geoff Bennington and Ian McLeod. Chicago: University of Chicago Press, 1987.

Dick, Steven J. and James E. Strick. *The Living Universe: NASA and the Development of Astrobiology*. New Brunswick, N.J.: Rutgers University Press, 2004.

DuPuy, Jean-Pierre. *The Mechanization of the Mind: On the Origins of Cognitive Science*. Translated by M. B. DeBevoise. Princeton, N.J.: Princeton University Press, 2000.

Duyfhuizen, Bernard. *Narratives of Transmission*. Rutherford, N.J.: Fairleigh Dickinson University Press, 1992.

Economides, Louise. "'Mont Blanc' and the Sublimity of Materiality." *Cultural Critique* 61 (Fall 2005): 87–114.

Esposito, Elena. "Two-Sided Forms in Language." In *Problems of Form*, edited by Baecker. 78–98.

The Fly. Dir. David Cronenberg. Twentieth-Century Fox, 1986.

The Fly. Dir. Kurt Neumann. Twentieth-Century Fox, 1958.

Foster, Thomas. "'The Sex Appeal of the Inorganic': Posthuman Narratives and the Construction of Desire." In *Centuries' Ends*, edited by Newman. 276–301.

Freeland, Cynthia A. "Feminist Frameworks for Horror Films." In *Post-Theory: Reconstructing Film Studies*, edited by David Bordwell and Noël Carroll. Madison: University of Wisconsin Press, 1996. 195–218

Genette, Gérard. *Narrative Discourse: An Essay on Method*. Translated by Jane E. Lewin. Ithaca, N.Y.: Cornell University Press, 1980.

Gerovitch, Slava. *From Newspeak to Cyberspeak: A History of Soviet Cybernetics*. Cambridge, Mass.: MIT Press, 2002.

Gold, Barri J. "Reproducing Empire: *Moreau* and Others." In *Nineteenth Century Studies* 14 (2000): 173–98.

Graham, Elaine L. *Representations of the Post/Human: Monsters, Aliens, and Others in Popular Culture*. New Brunswick, N.J.: Rutgers University Press, 2002.

Gray, John. *Straw Dogs: Thoughts on Humans and Other Animals*. London: Granta, 2002.

Grössing, Gerhard, Joseph Hartmann, Werner Korn, and Albert Müller, eds. *Heinz von Foerster 90*. Vienna: Edition Echoraum, 2001.

Gumbrecht, Hans Ulrich, and L. Ludwig Pfeiffer, eds. *Materialities of Communication*. Trans. William Whobrey. Stanford, Calif.: Stanford University Press, 1994.

Habermas, Jürgen. *The Philosophical Discourse of Modernity: Twelve Lectures*, translated by Frederick Lawrence. Cambridge, Mass.: MIT Press, 1987.

Halberstam, Judith and Ira Livingston, eds. *Posthuman Bodies*. Bloomington: Indiana University Press, 1995.

Halpern, Richard. "The Lyric in the Field of Information: Autopoiesis and History in Donne's *Songs and Sonnets.*" *Yale Journal of Criticism* 6, no. 1 (1993): 185–215.

Hansen, Mark B. N. "Cinema Beyond Cybernetics, or How to Frame the Digital Image." *Configurations* 10, no. 1 (2002): 51–90.

———. *New Philosophy for New Media.* Cambridge, Mass.: MIT Press, 2004.

Haraway, Donna J. "A Cyborg Manifesto: Science, Technology, and Socialist-Feminism in the Late Twentieth Century." In Haraway, *Simians, Cyborgs, and Women.* 149–81.

———. "Cyborgs and Symbionts: Living Together in the New World Order." In *The Cyborg Handbook*, edited by Chris Hables Gray with Steven Mentor and Heidi J. Figueroa-Sarriera. New York: Routledge, 1995. xi–xx.

———. *Modest_Witness@Second_Millennium. FemaleMan©_Meets_OncoMouse™: Feminism and Technoscience.* New York: Routledge, 1997.

———. *Primate Visions.* New York: Routledge, 1989.

———. "The Promises of Monsters: A Regenerative Politics of Inappropriate/d Others." In *Cultural Studies*, edited by Lawrence Grossberg, Cary Nelson, and Paula A. Treichler. New York: Routledge, 1992. 295–337.

———. *Simians, Cyborgs, and Women: The Reinvention of Nature.* New York: Routledge, 1991.

Hardison, O. B. *Disappearing Through the Skylight: Culture and Technology in the Twentieth Century.* New York: Viking, 1989.

Hayles, Katherine. "Chaos as Dialectic: Stanislaw Lem and the Space of Writing." In Hayles, *Chaos Bound.* 115–40.

———. *Chaos Bound: Orderly Disorder in Contemporary Literature and Science.* Ithaca, N.Y.: Cornell University Press, 1990.

———. *How We Became Posthuman: Virtual Bodies in Cybernetics, Literature, and Informatics.* Chicago: University of Chicago Press, 1999.

———. "Making the Cut: The Interplay of Narrative and System, or What Systems Theory Can't See." In *Observing Complexity*, edited by Rasch and Wolfe, 137–62.

Hellerstein, N. S. *Diamond: A Paradox Logic.* River Edge, N.J.: World Scientific, 1997.

Hendershot, Cyndy. "The Cold War Horror Film: Taboo and Transgression in *The Bad Seed*, *The Fly*, and *Psycho.*" *The Journal of Popular Film and Television* 29, no. 1 (Spring 2001): 21–31.

Herman, David, ed. *Narrative Theory and the Cognitive Sciences.* Stanford, Calif.: CSLI Publications, 2003.

Heylighen, Frances and Cliff Joslyn. "Cybernetics and Second-Order Cybernetics." In *Encyclopedia of Physical Science and Technology*, edited by R. A. Meyers. 3rd ed. New York: Academic Press, 2001.

Holden, Rebecca J. "The High Costs of Cyborg Survival: Octavia Butler's *Xenogenesis* Trilogy." *Foundation: The International Review of Science Fiction* 72 (Spring 1998): 49–56.

Hughes, Patrick and George Brecht. *Vicious Circles and Infinity: An Anthology of Paradoxes*. New York: Penguin, 1975.

Hutcheon, Linda. *Narcissistic Narrative: The Metafictional Paradox*. Waterloo, Ont.: Wilfred Laurier University Press, 1980.

Jacobs, Naomi. "Posthuman Bodies and Agency in Octavia Butler's *Xenogenesis*." In *Dark Horizons: Science Fiction and the Dystopian Imagination*, edited by Raffaella Baccolini and Tom Moylan. New York: Routledge, 2003. 91–111.

Jahn, Manfred. *Narratology: A Guide to the Theory of Narrative*. http://www.uni-koeln.de/~ame02/pppn.htm.

Jakobson, Roman. "Linguistics and Poetics." In *Style in Language*, edited by Thomas Sebeok. New York: John Wiley, 1960. 350–77.

Kauffman, Louis H. "Self-Reference and Recursive Forms." *Journal of Social and Biological Structure* 10 (1987): 53–72.

Keats, John. *Hyperion: A Fragment*. In *John Keats: Complete Poems*, edited by Jack Stillinger. Cambridge, Mass.: Harvard University Press, 1982.

Kirkman, Greg. "The Annotated Fly." http://annotatedfly1986.blogspot.com.

Kittler, Friedrich. *Gramophone, Film, Typewriter*. Trans. Geoffrey Winthrop-Young and Michael Wutz. Stanford, Calif.: Stanford University Press, 1999.

Knee, Adam. "The Metamorphosis of *The Fly*." *Wide Angle* 14, no. 1 (January 1992): 20–34.

Knight, Damon. *Beyond the Barrier*. Garden City, N.Y.: Doubleday, 1964.

Krumm, Pascale. "*The Island of Dr. Moreau*, or the Case of Devolution." *Foundation* 75 (Spring 1999): 51–62.

Kukla, André. *Social Constructivism and the Philosophy of Science*. New York: Routledge, 2000.

Kurzweil, Ray. *The Age of Spiritual Machines: When Computers Exceed Human Intelligence*. New York: Viking, 1999.

Langelaan, George. "The Fly." *Playboy* (June 1957): 17–18, 22, 36, 38, 46, 64–68.

———. "The Fly." In *The Playboy Book of Science Fiction and Fantasy*. Chicago: Playboy Press, 1966. 1–39.

———. *The Masks of War*. Garden City, N.Y.: Doubleday, 1959.

Lanier, Jaron. "One Half of a Manifesto." http://www.edge.org/3rd_culture/lanier/lanier_p1.html.

Latour, Bruno. *Aramis, or the Love of Technology*. Translated by Catherine Porter. Cambridge, Mass.: Harvard University Press, 1996.

———. *Pandora's Hope: Essays on the Reality of Science Studies*. Cambridge, Mass.: Harvard University Press, 1999.

———. "The Promises of Constructivism." In *Chasing Technoscience: Matrix for Materiality*, edited by Don Ihde and Evan Selinger. Bloomington: Indiana University Press, 2003. 27–46.

———. *Science in Action: How to Follow Scientists and Engineers through Society*. Cambridge, Mass.: Harvard University Press, 1987.

———. *We Have Never Been Modern*. Translated by Catherine Porter. Cambridge, Mass.: Harvard University Press, 1993.

Latour, Bruno, and Steve Woolgar. *Laboratory Life: The Construction of Scientific Facts*. 2nd ed. Princeton, N.J.: Princeton University Press, 1986.

Lem, Stanislaw. *The Cyberiad: Fables for the Cybernetic Age*. Translated by Michael Kandel. New York: Harvest, 1985.

———. *Summa Technologiae*. Kraków: Wydawnictwo Literackie, 1964.

———. "Thirty Years Later." In *A Stanislaw Lem Reader*, edited by Peter Swirski. Evanston, Ill.: Northwestern University Press, 1997. 67–91.

Lenoir, Timothy, ed. *Inscribing Science: Scientific Texts and the Materialities of Communication*. Stanford, Calif.: Stanford University Press, 1998.

———, ed. *Makeover: Writing the Body into the Posthuman Technoscape*. Part One: Embracing the Posthuman. *Configurations* 10, no. 2 (Spring 2002).

———, ed. *Makeover: Writing the Body into the Posthuman Technoscape*. Part Two: Corporeal Axiomatics. *Configurations* 10, no. 3 (Fall 2002).

Levinson, Marjorie. "What is New Formalism?" *PMLA* 122, no. 2 (March 2007): 558–69.

Livingston, Ira. *Between Science and Literature: An Introduction to Autopoetics*. Urbana: University of Illinois Press, 2006.

Luhmann, Niklas. *Art as a Social System*. Trans. Eva Knodt. Stanford, Calif.: Stanford University Press, 2000.

———. "The Autopoiesis of Social Systems." In *Sociocybernetic Paradoxes: Observation, Control and Evolution of Self-Steering Systems*, edited by Felix Geyer and Johannes van der Zouwen. London: Sage, 1986. 172–92.

———. "The Cognitive Program of Constructivism and a Reality That Remains Unknown." In *Theories*, by Luhmann, edited by Rasch. 128–52.

———. "The Control of Intransparency." In *Systems Research and Behavioral Science* 14 (1997): 359–71.

———. "Identity—What or How?" In *Theories*, by Luhmann, edited by Rasch. 113–27.

———. "I See Something You Don't See." In *Theories*, by Luhmann, edited by Rasch. 187–93.

———. *Love as Passion: The Codification of Intimacy*. Translated by Jeremy Gaines and Doris L. Jones. Stanford, Calif.: Stanford University Press, 1998.

———. "The Medium of Art." In *Essays on Self-Reference*, by Niklas Luhmann. New York: Columbia University Press, 1990. 215–26.

———. *Observations on Modernity*. Translated by William Whobrey. Stanford, Calif.: Stanford University Press, 1998.

———. "The Paradox of Form." In *Problems of Form*, edited by Baecker. 15–26.

———. "The Paradox of Observing Systems." In *Theories*, by Luhmann, edited by Rasch. 79–93.

———. *The Reality of the Mass Media*. Translated by Kathleen Cross. Stanford, Calif.: Stanford University Press, 2000.

———. *Social Systems*. Translated by John Bednarz Jr. with Dirk Baecker. Stanford, Calif.: Stanford University Press, 1995.

———. *Theories of Distinction: Redescribing the Descriptions of Modernity*. Edited by William Rasch. Stanford, Calif.: Stanford University Press, 2002.

———. "The World Society as a Social System." *International Journal of General Systems* 8 (1982): 131–38.

———. "What Is Communication?" In *Theories*, by Luhmann, edited by Rasch. 155–68.

Lyotard, Jean-François. *The Postmodern Condition: A Report on Knowledge*. Translated by Geoff Bennington and Brian Massumi. Minneapolis: University of Minnesota Press, 1984.

Margulis, Lynn. "Gaia Is a Tough Bitch." In *The Third Culture: Beyond the Scientific Revolution*, edited by John Brockman. New York: Touchstone, 1995. 129–46.

Margulis, Lynn, and Dorion Sagan. *Acquiring Genomes: A Theory of the Origins of Species*. New York: Basic Books, 2002.

———. *Microcosmos: Four Billion Years of Evolution from Our Microbial Ancestors*. New York: Summit Books, 1986.

———. *Microcosmos: Four Billion Years of Microbial Evolution*. Berkeley: University of California Press, 1997.

———. *What Is Life?* New York: Simon and Schuster, 1995.

Markley, Robert. *Dying Planet: Mars in Science and the Imagination*. Durham, N.C.: Duke University Press, 2005.

Maturana, Humberto and Francisco Varela. *The Tree of Knowledge: The Biological Roots of Human Understanding*. Rev. ed. Boston: Shambhala, 1998.

Max, D. T. "The Literary Darwinists." *New York Times Magazine*, November 4, 2005. 74–79.

McDermott, Jeanne. "Lynn Margulis: Vindicated Heretic." In *Gaia*, edited by Barlow. 47–56.

McMurry, Andrew. *Environmental Renaissance: Emerson, Thoreau, and the American System of Nature*. Athens: University of Georgia Press, 2003.

Mehaffy, Marilyn and AnaLouise Keating. "'Radio Imagination': Octavia Butler on the Poetics of Narrative Embodiment." *MELUS* 26, no. 1 (2001): 45–76.

Merrell, Floyd. *Semiotic Foundations: Steps Toward an Epistemology of Written Texts*. Bloomington: Indiana University Press, 1982.

———. *Unthinking Thinking: Jorge Luis Borges, Mathematics, and the New Physics*. West Lafayette, Ind.: Purdue University Press, 1991.

Michaels, Walter Benn. "Political Science Fictions." *NLH* 31 (2000): 649–64.

Mingers, John. *Self-Producing Systems: Implications and Applications of Autopoiesis*. New York: Plenum Press, 1995.

Moeller, Hans-Georg. *Luhmann Explained: From Souls to Systems*. Chicago: Open Court, 2006.

Montrose, Louis. "*A Midsummer Night's Dream* and the Shaping Fantasies of Elizabethan Culture: Gender, Power, Form." In *Rewriting the Renaissance: The Discourses of Sexual Difference in Early Modern Europe*, edited by Margaret W. Ferguson, Maureen Quilligan, and Nancy J. Vickers. Chicago: University of Chicago Press, 1986. 65–87.

Nelles, William. *Frameworks: Narrative Levels and Embedded Narrative*. New York: Peter Lang, 1997.

Newman, Beth. "Narratives of Seduction and the Seductions of Narrative: The Frame Structure of *Frankenstein*." *ELH* 53, no. 1 (Spring 1986): 141–63.

Newman, Robert, ed. *Centuries' Ends, Narrative Means*. Stanford, Calif.: Stanford University Press, 1996.

Nünning, Ansgar. "On the Perspective Structure of Narrative Texts: Steps Toward a Constructivist Narratology." In *Narrative Perspective*, edited by van Peer and Chatman. 207–23.

Otis, Laura. *Organic Memory: History and the Body in the Late Nineteenth and Early Twentieth Centuries*. Lincoln: Nebraska University Press, 1994.

Parisi, Luciana. *Abstract Sex: Philosophy, Bio-Technology and the Mutations of Desire*. New York: Continuum, 2004.

———. "Essence and Virtuality: The Incorporeal Desire of Lilith." *Anglistica* 4, no. 1 (2000): 191–212.

Paulson, William R. "The Literary Canon in the Age of its Technological Obsolescence." In *Reading Matters*, edited by Tabbi and Wutz. 227–49.

———. *The Noise of Culture: Literary Texts in a World of Information*. Ithaca, N.Y.: Cornell University Press, 1988.

Paxson, James J. "Kepler's Allegory of Containment, the Making of Modern Astronomy, and the Semiotics of Mathematical Thought." *Intertexts* 3, no. 2 (Fall 1999): 105–23.

Pay, Rex, ed. "Dreams." http://www.humanistictexts.org/chuang.htm. Adapted from *Chuang Tzu: Mystic, Moralist, and Social Reformer*, translated by Herbert A. Giles. London: Bernard Quaritch, 1926.

Pearce, Richard. "Enter the Frame." In *Surfiction: Fiction Now and Tomorrow*, edited by Raymond Federman. Chicago: Swallow Press, 1975. 47–57.

Peppers, Cathy. "Dialogic Origins and Alien Identities in Butler's Xenogenesis." *Science-Fiction Studies* 22 (1995): 47–63.

Peters, John Durham. *Speaking into the Air: A History of the Idea of Communication*. Chicago: University of Chicago Press, 1999.

Pharr, Mary Ferguson. "From Pathos to Tragedy: The Two Versions of *The Fly*." *Journal of the Fantastic in the Arts* 2, no. 1 (Spring 1989): 37–46.

Phelan, James. "Narrative Theory, 1966–2006: A Narrative." In *The Nature of Narrative*, edited by Robert Scholes, James Phelan, and Robert Kellogg. Rev. and expanded edition. New York: Oxford University Press, 2006. 283–336.

———. "Why Narrators Can Be Focalizers—And Why It Matters." In *Narrative Perspective*, edited by van Peer and Chatman. 51–64.

Prince, Gerald. "A Point of View on Point of View or Refocusing Focalization." In *Narrative Perspective*, edited by van Peer and Chatman. 43–50.

Pynchon, Thomas. *The Crying of Lot 49.* New York: Harper and Row, 1990.

Rasch, William. *Niklas Luhmann's Modernity: The Paradoxes of Differentiation.* Stanford, Calif.: Stanford University Press, 2000.

Rasch, William and Cary Wolfe, eds. *Observing Complexity: Systems Theory and Postmodernity.* Minneapolis: University of Minnesota Press, 2000.

Reinfandt, Cristoph, ed. "Systems Theory and Literature." Special issue, *European Journal of English Studies* 5, no. 3 (December 2001).

Robbins, Helen W. "'More Human Than I Am Alone': Womb Envy in David Cronenberg's *The Fly* and *Dead Ringers.*" In *Screening the Male: Exploring Masculinities in Hollywood Cinema*, edited by Steven Cohan and Ira Rae Hark. New York: Routledge, 1993. 134–47.

Roberts, David. "Self-Reference in Literature." In *Problems of Form*, edited by Baecker. 27–45.

Rossini, Manuela. "Figurations of Posthumanity in Contemporary Science/Fiction: All Too Human(ist)?" *Revista Canaria de Estudios Ingleses* 50 (April 2005): 21–36.

Roth, Marty. "Twice Two: *The Fly* and *Invasion of the Body Snatchers.*" In *Dead Ringers: The Remake in Theory and Practice*, edited by Jennifer Forrest and Leonard Koos. Albany, N.Y.: State University of New York Press, 2002. 225–41.

Rowell, Charles H. "An Interview with Octavia E. Butler." In *Callaloo* 20, no. 1 (1997): 47–66.

Royle, Nicholas. *Telepathy and Literature: Essays on the Reading Mind.* Cambridge, Mass.: Basil Blackwell, 1991.

Ryan, Marie-Laure. *Avatars of Story.* Minneapolis: University of Minnesota Press, 2006.

Sagan, Carl. *Contact.* New York: Pocket Books, 1986.

Sagan, Dorion, and Lynn Margulis. "The Uncut Self." In *Dazzle Gradually: Reflections on the Nature of Nature*, by Dorion Sagan and Lynn Margulis. White River Junction, Vt.: Chelsea Green, 2007. 16–25.

Schiltz, Michael. "Space is the Place: The *Laws of Form* and Social Systems." *Thesis Eleven* 88 (February 2007): 8–30; rp. in *Emergence and Embodiment*, edited by Clarke and Hansen.

Schmidt, Siegfried. "Literary Systems as Self-Organizing Systems." In *Selforganization: Portrait of a Scientific Revolution*, edited by Wolfgang Krohn, Günter Küpper, and Helga Nowotny. Boston: Kluwer, 1990. 143–53.

Schwanitz, Dietrich. "Systems Theory and the Difference between Communication and Consciousness: An Introduction to a Problem and Its Context." *MLN* 111 (1996): 488–505.

Serres, Michel. *Genesis.* Translated by Geneviève James and James Nielson. Ann Arbor: University of Michigan Press, 1995.

———. *Hermes: Literature, Science, Philosophy.* Edited by Josué V. Harari and David F. Bell. Baltimore: Johns Hopkins University Press, 1982.

———. *The Parasite.* Translated by Lawrence R. Schehr. Intro. Cary Wolfe. Minneapolis: University of Minnesota Press, 2007.

Serres, Michel, with Bruno Latour. *Conversations on Science, Culture, and Time.* Translated by Roxanne Lapidus. Ann Arbor: University of Michigan Press, 1995.

Shakespeare, William. *A Midsummer Night's Dream.* In *Complete Works*, edited by Bevington. 148–79.

Slonczewski, Joan. "Octavia Butler's *Xenogenesis* Trilogy: A Biologist's Response." http://biology.kenyon.edu/slonc/books/butler1.html.

Smith, Stephanie A. "Morphing, Materialism, and the Marketing of *Xenogenesis*." *Genders* 18 (Winter 1993): 67–86.

Smithson, Robert. "Entropy and the New Monuments." In *Robert Smithson: The Collected Writings.* Edited by Jack Flam. Berkeley: University of California Press, 1996. 10–23.

Sonea, Sorin, and Leo G. Mathieu. *Prokaryotology: A Coherent View.* Montréal, Québec: Les Presses de l'Université de Montréal, 2000.

Spencer-Brown, George. *Laws of Form.* 1969; New York: E. P. Dutton, 1979.

Squier, Susan. "From Omega to Mr. Adam: The Importance of Literature for Feminist Science Studies." *Science, Technology, and Human Values* 24, no. 1 (Winter 1999): 131–57.

———. "Interspecies Reproduction: Xenogenic Desire and the Feminist Implication of Hybrids." *Cultural Studies* 12, no. 1 (1998): 360–81.

Stanzel, F. K. *A Theory of Narrative.* Translated by Charlotte Goedsche. New York: Cambridge University Press, 1984.

Sterling, Bruce. *Schismatrix.* 1985. In *Schismatrix Plus*, by Sterling. New York: Ace Books, 1996. 3–236.

Stites, Richard. *Revolutionary Dreams: Utopian Vision and Experimental Life in the Russian Revolution.* New York: Oxford University Press, 1989.

Tabbi, Joseph. *Cognitive Fictions.* Minneapolis: University of Minnesota Press, 2002.

Tabbi, Joseph, and Michael Wutz, eds. *Reading Matters: Narratives in the New Media Ecology.* Ithaca, N.Y.: Cornell University Press, 1997.

Telotte, J. P. "Crossing Genre Boundaries/Bound by Fantasy." In *Science Fiction Film*, edited by J. P. Telotte. New York: Cambridge University Press, 2001. 179–95.

Todorov, Tzvetan. *The Poetics of Prose.* Translated by Richard Howard. Ithaca, N.Y.: Cornell University Press, 1977.

Udwin, Victor. "Autopoiesis and Poetics." In *Textuality and Subjectivity: Essays on Language and Being*, edited by Eitel Timm, Kenneth Mendoza, and Dale Gowan. Columbia, S.C.: Camden House, 1991. 1–13.

van Peer, Willie, and Seymour Chatman, eds. *New Perspectives on Narrative Perspective.* Albany, N.Y.: SUNY Press, 2001.

bibliography

234

234

234

234

234

234

Varela, Francisco J. *Principles of Biological Autonomy*. New York: North Holland, 1979.

Venn, Couze and Mike Featherstone. "Modernity." *Theory, Culture and Society* 23, nos. 2–3 (2006): 457–65.

von Foerster, Heinz. "For Niklas Luhmann: 'How Recursive is Communication?'" In *Understanding Understanding*, by von Foerster. 305–23.

———. "Interview." With Stefano Franchi, Güven Güzeldere, and Eric Minch. *Stanford Electronic Humanities Review* 4, no. 2. http://www.stanford.edu/group/SHR/4-2/text/interviewvonf.html.

———. "Laws of Form." In *The Last Whole Earth Catalog*, edited by Stewart Brand. Palo Alto, Calif.: Portola Institute, 1971. 14.

———. "Objects: Tokens for (Eigen-)Behaviors." In *Understanding Understanding*, by von Foerster. 261–71.

———. "On Self-Organizing Systems and Their Environments." In *Understanding Understanding*, by von Foerster. 1–19.

———. "Perception of the Future and the Future of Perception." In *Understanding Understanding*, by von Foerster. 199–210.

———. *Understanding Understanding: Essays on Cybernetics and Cognition*. New York: Springer, 2003.

von Foerster, Heinz, with Bernhard Poerksen. *Understanding Systems: Conversations on Epistemology and Ethics*. New York: Kluwer Academic/Plenum, 2002.

Waugh, Patricia. *Metafiction: The Theory and Practice of Self-Conscious Fiction*. New York: Routledge, 1984.

Weaver, Warren. "Recent Contributions to the Mathematical Theory of Communication." In *The Mathematical Theory of Communication*, by Warren Weaver and Claude Shannon. Urbana: University of Illinois Press, 1949. 3–28.

Weinstone, Ann. *Avatar Bodies: A Tantra for Posthumanism*. Minneapolis: University of Minnesota Press, 2004.

Weiskel, Thomas. *The Romantic Sublime: Studies in the Structure and Psychology of Transcendence*. Baltimore: Johns Hopkins University Press, 1976.

Wellbery, David E. "Redescription: Literary Semiotics, Deconstruction, Systems Theory." Unpublished manuscript.

———, ed. "Observation, Difference, Form: Literary Studies and Second-Order Cybernetics." *MLN* 111, no. 3 (April 1996).

Wells, H. G. *The Island of Dr. Moreau*. New York: Dover, 1996.

———. *The War of the Worlds*. New York: Signet Classic, 1986.

Wesling, Donald. "Michel Serres, Bruno Latour, and the Edges of Historical Periods." *Clio* 26, no. 2 (1997). 189–204.

White, Eric. "The Erotics of Becoming: *Xenogenesis* and 'The Thing.'" *Science-Fiction Studies* 20, no. 3 (November 1993): 394–408.

Whitman, Walt. "Out of the Cradle Endlessly Rocking." In *Leaves of Grass and Selected Poems*, by Whitman, edited by Lawrence Buell. New York: Modern Library, 1981.

Wicke, Jennifer. "Fin de Siècle and the Technological Sublime." In *Centuries' Ends*, edited by Newman. 302–15.

Wiener, Norbert. *The Human Use of Human Beings: Cybernetics and Society.* Boston: Houghton Mifflin, 1950.

Wilden, Anthony. *System and Structure: Essays in Communication and Exchange.* 2nd ed. London: Tavistock, 1980.

Williamson, Donald I. *The Origins of Larvae.* Boston: Kluwer, 2003.

Winthrop-Young, Geoffrey, and Michael Wutz. "Translators' Introduction: Friedrich Kittler and Media Discourse Analysis." In *Gramophone*, by Kittler, xi–xxxviii.

Wolfe, Cary. "In the Shadow of Wittgenstein's Lion: Language, Ethics, and the Question of the Animal." In *Zoontologies: The Question of the Animal*, edited by Cary Wolfe. Minneapolis: University of Minnesota Press, 2003. 1–57.

———. "Meaning as Event-Machine, or, Systems Theory and 'The Reconstruction of Deconstruction.'" In *Emergence and Embodiment*, edited by Clarke and Hansen.

———. "Systems Theory: Maturana and Varela with Luhmann." In *Critical Environments: Postmodern Theory and the Pragmatics of the "Outside,"* by Wolfe. Minneapolis: University of Minnesota Press, 1998. 40–83.

Yu, Jeboon. "The Representation of Inappropriate/d Others: The Epistemology of Donna Haraway's Cyborg Feminism and Octavia Butler's *Xenogenesis* Series." *Journal of English Language and Literature* 50, no. 3 (2004): 759–77.

Zeilinger, Anton. "Quantum Teleportation." *Scientific American* (April 2000): 50–59.